Vugar Mehrabov

On a spectrum of Pauli operator with periodic coefficients

Vugar Mehrabov

On a spectrum of Pauli operator with periodic coefficients

LAP LAMBERT Academic Publishing

Impressum / Imprint

Bibliografische Information der Deutschen Nationalbibliothek: Die Deutsche Nationalbibliothek verzeichnet diese Publikation in der Deutschen Nationalbibliografie; detaillierte bibliografische Daten sind im Internet über http://dnb.d-nb.de abrufbar.
Alle in diesem Buch genannten Marken und Produktnamen unterliegen warenzeichen-, marken- oder patentrechtlichem Schutz bzw. sind Warenzeichen oder eingetragene Warenzeichen der jeweiligen Inhaber. Die Wiedergabe von Marken, Produktnamen, Gebrauchsnamen, Handelsnamen, Warenbezeichnungen u.s.w. in diesem Werk berechtigt auch ohne besondere Kennzeichnung nicht zu der Annahme, dass solche Namen im Sinne der Warenzeichen- und Markenschutzgesetzgebung als frei zu betrachten wären und daher von jedermann benutzt werden dürften.

Bibliographic information published by the Deutsche Nationalbibliothek: The Deutsche Nationalbibliothek lists this publication in the Deutsche Nationalbibliografie; detailed bibliographic data are available in the Internet at http://dnb.d-nb.de.
Any brand names and product names mentioned in this book are subject to trademark, brand or patent protection and are trademarks or registered trademarks of their respective holders. The use of brand names, product names, common names, trade names, product descriptions etc. even without a particular marking in this work is in no way to be construed to mean that such names may be regarded as unrestricted in respect of trademark and brand protection legislation and could thus be used by anyone.

Coverbild / Cover image: www.ingimage.com

Verlag / Publisher:
LAP LAMBERT Academic Publishing
ist ein Imprint der / is a trademark of
OmniScriptum GmbH & Co. KG
Heinrich-Böcking-Str. 6-8, 66121 Saarbrücken, Deutschland / Germany
Email: info@lap-publishing.com

Herstellung: siehe letzte Seite /
Printed at: see last page
ISBN: 978-3-659-69483-7

V.A. MEHRABOV

On a spectrum of Pauli operator

with periodic coefficients

Contents

INTRODUCTION

Periodic differential operators are one of the most important classes of operators studied in the spectral theory $[9,10,24,27]$. Investigation of the spectrum of the periodic Schrödinger operator (Hill operator) with real potential was begun in the classical works $[18,28]$ in connection with the problem of stability in celestial mechanics, since the closure of the stability zones of the Hill equation coincides with the spectrum of the corresponding operator. As for the multidimensional Schrödinger operator

$$H = -\Delta + q(x), \quad x \in R^n \tag{1}$$

with periodic (with respect to an arbitrary lattice Ω) potential $q(x)$, its spectral theory is the mathematical basis of quantum theory of solids. As explained in the pioneering work of physicists, even the simplest study of the spectrum of three-dimensional periodic Schrödinger operator H allows one to explain the nature of the physical properties of real crystals $[1-5,17,28]$, but to the study of this problem on a purely mathematical level very few works $[11,30,31,32]$ have been devoted. In the work by I.M. Gelfand $[11]$ spectral decomposition for self-adjoint differential operators with periodic potentials is given and is proved that its spectrum $S(H)$ of the operator H coincides with the union of the spectrum $S(H_t)$ of operators H_t generated by (1) in the fundamental domain R^n/Ω (parallelepiped) of the lattice Ω with quasi-periodic boundary conditions with quasi-momentum $t \in R^n$

$$S(H) = \bigcup_{t \in R^n} S(H_t) \tag{2}$$

This relationship between the spectra of periodic operators and regular differential operators opens up opportunities for the use of asymptotic formulas obtained for the eigenvalues of regular operators studied in $[2,5]$ for the multidimensional periodic Schrödinger operator.

In this monograph similar problems are considered for the periodic Pauli operator, i.e. the relation (2) is established for the Pauli operator, and for the first time asymptotic formulas are obtained for the eigenvalues of this operator in the parallelogram and for some series of the eigenvalues in the parallelepiped.

We consider two and three dimensional Pauli operator $H_t(a, V(h))$ generated in $L_2(F) \times L_2(F)$ by the expression

$$H(a, V(h)) = \left((-i\nabla - a)^2 + V(h) \right) \cdot I + \sigma \cdot B \tag{3}$$

and boundary conditions

$$u\left(h+\omega_j\right)=e^{2\pi i t_j}u(h),\quad h\in R^3,\tag{4}$$

where $V(x)$ is periodic relatively the lattice $\Omega=\{m_1\omega_1+m_2\omega_2+m_3\omega_3:m_1,m_2,m_3\in Z\}$ and enough smooth function; $\Gamma=\{n_1\gamma^1+n_2\gamma^2+n_3\gamma^3:n_1,n_2,n_3\in Z\}$ is dual to Ω lattice i.e. $\left(\gamma^i,\omega_j\right)=2\pi\delta_{ij}$; $t=t_1\gamma^1+t_2\gamma^2+t_3\gamma^3$; $B=[\nabla,a]$ is a magnetic field generated by the vector potential $a=\left(a_1,a_2,a_3\right)$, i.e. $B=\left(B_1,B_2,B_3\right)$, where

$$B_1=\begin{vmatrix}\dfrac{\partial}{\partial y}&\dfrac{\partial}{\partial z}\\a_2&a_3\end{vmatrix},\quad B_2=\begin{vmatrix}\dfrac{\partial}{\partial z}&\dfrac{\partial}{\partial x}\\a_3&a_1\end{vmatrix},\quad B_3=\begin{vmatrix}\dfrac{\partial}{\partial x}&\dfrac{\partial}{\partial y}\\a_1&a_2\end{vmatrix};$$

$\sigma=\left(\sigma_1,\sigma_2,\sigma_3\right)$ where

$$\sigma_1=\begin{pmatrix}0&1\\1&0\end{pmatrix},\quad \sigma_2=\begin{pmatrix}0&-i\\i&0\end{pmatrix},\quad \sigma_3=\begin{pmatrix}1&0\\0&-1\end{pmatrix};\quad I=\begin{pmatrix}1&0\\0&1\end{pmatrix},$$

F is a fundamental domain of some lattice Ω - i.e. a parallelepiped; F^* is a fundamental domain of the lattice Γ.

In two dimensional case we have

$$\sigma=\begin{pmatrix}1&0\\0&-1\end{pmatrix},\quad B=[\nabla,a]=\frac{\partial}{\partial x_1}a_2-\frac{\partial}{\partial x_2}a_1$$

i.e.

$$H_0(a)=\left(-i\nabla-a\right)^2\cdot I+\sigma\cdot B.\tag{3'}$$

It is known that Pauli operator is of one of the important operators of the quantum physics. It describes the motion of the particle with spin in the electromagnetic field. In addition to the mechanical and magnetic moment produced by the motion of the center of gravity of the electron, the electron must be attributed to the intrinsic mechanical and magnetic moments, as if it was not a material point, and a rotating charged top. This mechanical and magnetic moment is called a spin, and the phenomenon is called as a spin of the electron [3,16]. Mechanical and magnetic moments produced by the motion of the center of gravity of the electron are called orbital.

The aim of this work is obtaining of the asymptotical formulas for the eigenvalues of the Pauli operator $H_t\left(a,V(h)\right)$ in parallelogram and parallelepiped.

A big number of works are devoted to the perturbation theory of the Laplace operator with potential $q(x)$ [14,15,19,25,29]. Usually the operator

$$T_F\left(\varepsilon\right)=-\Delta+\varepsilon q(x).$$

is considered in those works. If for the eigenvalues $\Lambda_n(\varepsilon)$ and eigenfunctions $\Psi_n(x,\varepsilon)$ of this operator to write formal Ralley-Schrodinger series

$$\Lambda_n(\varepsilon) = \lambda_n + \varepsilon\lambda_n^1 + \varepsilon^2\lambda_n^2 + ...$$
$$\Psi_n(x,\varepsilon) = \varphi_n(x) + \varepsilon\varphi_n^1(x) + \varepsilon^2\varphi_n^2(x) + ...$$

(5)

then it is necessary to take ε less that $\min_{k \neq n}|\lambda_n - \lambda_k|$ for their convergence. But in multidimensional case $\lim_{n \to \infty}\left(\min_{k \neq n}|\lambda_n - \lambda_k|\right) = 0$. Therefore the ε, for which the series (5) converge for any n does not exists.

Note that there are works in which the asymptotics series of the eigenvalues of the Laplace operator $[17]$, as well as numerous works on the asymptotic behavior of $N(\lambda)$ (the distribution of eigenvalues). There are also numerous works on semi classical asymptotics $[11,19]$ (i.e. when the small parameter stands at the Laplace operator) of the Schrödinger operator and other elliptic operators.

Here we use the method proposed O.A. Veliev $[6-8]$, where the asymptotic formulas are obtained for the eigenvalues and eigenfunctions of the Schrödinger operator in parallelepiped.

Here we give some definitions and results from those studies, which will be used.

Definition. Eigenvalue λ_k of the operator A is called related to the eigenvalues λ_n through perturbation B if

$$\left|(B\varphi_k, \varphi_n)\right| > \lambda_k^{-\alpha},$$

where $\alpha = const$, $0 < \alpha < 1$, and φ_k is an eigenfunction of the operator A corresponding to the eigenvalue λ_k.

If there exist the indexes $k_1, k_2, ..., k_p$ $(p \leq m-1)$ such that

$$\left|(B\varphi_k, \varphi_{k_1})\right| > \lambda_k^{-\alpha m}, \left|(B\varphi_{k_1}, \varphi_{k_2})\right| > \lambda_k^{-\alpha m}, ..., \left|(B\varphi_{k_{p-1}}, \varphi_{k_p})\right| > \lambda_k^{-\alpha m}, \left|(B\varphi_{k_p}, \varphi_n)\right| > \lambda_k^{-\alpha m},$$

then we say that the eigenvalue λ_k of the operator A associated with the eigenvalue λ_n through perturbation $B^{(m)}$.

Large enough eigenvalue λ_k is called **_non-resonance_** (**_m-order non-resonance_**) if associated with it through B ($B^{(m)}$) eigenvalue λ_n satisfies to the estimation $|\lambda_k - \lambda_n| > \lambda_k^{\alpha}$.

The rest of the eigenvalues are called **_resonance_**. The set R of the eigenvalues is called an **_isolated resonance set_** if any eigenvalue $\lambda_n \notin R$ associated with at least one eigenvalue from R (

$\lambda \in D_m(\lambda_k)$ (where $D_m(\lambda_k)$ is the set of eigenvalues associated with the eigenvalue $\lambda_k \in R$ through $B^{(m)}$) are at a distance from R more that $(\rho(0,R))^\alpha$, where

$$\rho(\lambda, R) = \min_{k:\lambda_k \in R} |\lambda - \lambda_k|.$$

Other words

$$\rho(\lambda_n, R) > (\rho(0,R))^\alpha.$$

The eigenvalue λ_k is called non-resonance if the rest of eigenvalues λ_n are not associated or are far from it.

Eigenvalue Λ_N of the operator $A + B$ (where A is unbounded self-adjoint operator, operator B is bounded) denote by $\Lambda_N(\lambda_k)$ (or simply $\Lambda(\lambda_k)$), if

$$\left| (\Psi_N, \varphi_k) \right| = \max_{s=1,2,\dots} \left| (\Psi_N, \varphi_s) \right|.$$

Here Ψ_N is an eigenfunction of the operator $A + B$ corresponding to the eigenvalue Λ_N.

If the max is reached for few s the arbitrary of them may be chosen. We regard that $\Lambda_N(\lambda_k)$ corresponds to λ_k.

As it is proved in $[6]$ the set of eigenvalues $\{\lambda_k : k = 1,2,\dots\}$ of the Laplace operator consists of two disjoint subsets characterized as a set of m-order ($m = 1,2,\dots$) non-resonance eigenvalues and the set of resonance eigenvalues. The elements of these sets under perturbation of periodic, smooth potential are perturbed by certain way for each set. Further, it is proved that in some sense the main part of the eigenvalues $\Lambda_1, \Lambda_2, \dots, \Lambda_N, \dots$ of the Schrödinger operator T_F (where F is one of the sets: parallelepiped, sphere-circle, cylinder) is in the $O(\lambda_k^{-\alpha m})$ neighborhood of the points

$$\lambda_k + F_0(\lambda_k) + F_1(\lambda_k) + \dots + F_{m-1}(\lambda_k), \tag{6}$$

where λ_k is an eigenvalue of the Laplace operator T_F^0 and is m-order non-resonance, α is a positive number, $F_n(\lambda_k)$ is explicitly expressed by λ_k and potential $q(x)$ and has the order $O(\lambda_k^{-\alpha m})$. The rest of eigenvalues of the Schrodinger operator are in the $O(\tilde{\lambda}_k^{-\alpha})$ neighborhood of the eigenvalues $\tilde{\lambda}_k$ of some Sturm–Liouville operators or matrices that are explicitly constructed.

Now we give a detail description of the work.

The book consists of introduction, two chapters and references

In the **first chapter of the first chapter** using the Rid-Saymon method $[26]$ we expand the periodic Pauli operator in layers and prove the formula (2) for this operator. We use the following theorem from[26]. (Here the numbering of formulas and theorems corresponds to the numbering in the first and the second chapters of the monograph).

Theorem 1.1.1. Let $A = \int_M^{\oplus} A(m)d\mu(m)$, where $A(\cdot)$ is measurable and $A(m)$ is self-adjoint operator for any m. Then $\lambda \in \sigma(A)$ if and only if, when for any $\varepsilon > 0$

$$\mu\left(\{m \in M \mid \sigma(A(m)) \cap (\lambda - \varepsilon, \lambda + \varepsilon) \neq \emptyset\}\right) > 0. \tag{1.1.7}$$

In our case $A \equiv H(a, V(h))$ is generated in $L_2(R^2) \times L_2(R^2)$ by the expression (3), $A(m) \equiv H_t(a, V(h))$ is generated in $L_2(F) \times L_2(F)$ by the expression (3) and boundary conditions (4), $M \equiv F^*$ is a fundamental domain of the lattice Γ that is dual to the lattice Ω.

Using the (A) type analyticity of the operator $H_t(a, V(h))$ and the above theorem we prove the following theorem

Theorem 1.1.2.

$$\sigma(H(a, V(h))) = \bigcup_{t \in F} \sigma(H_t(a, V(h))). \tag{1.1.21}$$

Here $\sigma(H)$ is a spectrum of the operator H.

Finding the asymptotics of the eigenvalues of the operator $H_t(a, V(h))$ in principle we find asymptotic formulas for generalized (in some sense) eigenvalues of the operator $H_t(a, V(h))$. Therefore, the future, we will analyze only the operator $H_t(a, V(h))$.

In the **second paragraph of the first chapter**, we obtain asymptotic formulas for the eigenvalues of the two-dimensional $H_t(a, V(x))$ operator in the parallelogram. Denote by $H_0^t(a)$ the operator generated in $L_2(F) \times L_2(F)$ by the expression (3') and the boundary conditions (4). By the elementary computations we find that the eigenvalues and eigenfunctions of the operator $H_0^t(a)$ are as follows

$$\phi_\gamma(x) = \begin{pmatrix} 0 \\ e^{i(\gamma + t, x)} \end{pmatrix}, \quad \text{where} \quad \gamma \in \Gamma, \ t \in F, \tag{1.2.6}$$

$$\lambda_\gamma = |\gamma + t|^2 - 2(\gamma + t, a) + a^2 - i[\gamma + t, a], \tag{1.2.7}$$

$$\phi^\gamma(x) = \begin{pmatrix} e^{i(\gamma + t, x)} \\ 0 \end{pmatrix}, \quad \text{where} \quad \gamma \in \Gamma, \ t \in F, \tag{1.2.8}$$

$$\lambda^\gamma = |\gamma + t|^2 - 2(\gamma + t, a) + a^2 + i[\gamma + t, a]. \tag{1.2.9}$$

For the sake of simplicity first assume that the potential $V(x)$ is a trigonometric polynomial

$$V(x) = \sum_{\gamma \in Q} V_\gamma e^{i(\gamma, x)},$$ (1.2.10)

where $\qquad V_\gamma = \int_F V(x) e^{-i(\gamma, x)} dx, \qquad Q = \{\gamma \in \Gamma : V_\gamma \neq 0\}.$ (1.2.11)

Let

$$V_\gamma(\alpha) = \left\{ x \in R^2 : \left| |x|^2 - |x + \gamma|^2 \right| < |x|^\alpha \right\},$$ (1.2.15)

where $\gamma \in Q^{(m)}, \qquad Q^{(m)} = \{ \gamma : \gamma = \gamma_1 + \gamma_2 + \ldots + \gamma_p, \gamma_1, \gamma_2, \ldots, \gamma_p \in Q; p \leq m \}.$ (1.2.16)

If $\gamma + t \notin \bigcup_{\gamma_1 \in Q^m} (V_{\gamma_1}(\alpha)) \equiv V^{(m)}(\alpha)$, then eigenvalues λ_γ and λ^γ of the operator $H_0^t(a)$ we call

m-order non-resonance. The set $V_\gamma(\alpha)$ then is called a resonance set. It is obvious that non-resonant eigenvalues constitute the bulk of all eigenvalues of the operator $H_0^t(a)$.

Let us numerate all eigenvalues of the operator $H_t(a, V(x))$ due the pints of the lattice Γ i.e. denote them by $\Lambda(\lambda_\gamma)$ or $\Lambda(\lambda^\gamma)$, where $\gamma \in \Gamma$. The set $\{\Lambda(\lambda_\gamma), \Lambda(\lambda^\gamma) : \gamma \in \Gamma\}$ constitutes all spectrum of the operator $H_t(a, V(x))$. The following theorems are proved.

Theorem 1.2.2. a) If λ_γ is m-order non-resonance (i.e. $\gamma + t \notin V^{(m)}(\alpha)$), then corresponding eigenvalue $\Lambda(\lambda_\gamma)$ of the operator $H_t(a, V(x))$ satisfies to the asymptotics (1.2.40)

$$\Lambda(\lambda_\gamma) = \lambda_\gamma + \tilde{F}_{m-1}(\lambda_\gamma) + O\left(|\gamma + t|^{-\alpha m}\right),$$ (1.2.20)

where $\qquad F_1(\lambda_\gamma) = \tilde{F}_1(\lambda_\gamma) = S_1(\lambda_\gamma) = \sum_{\gamma_1} \frac{|V_{\gamma - \gamma_1}|}{\lambda_\gamma - \lambda_{\gamma_1}},$

$$\tilde{F}_n(\lambda_\gamma) = \sum_{k=1}^n S_k\left(\tilde{F}_{m-1}(\lambda_\gamma), \lambda_\gamma\right), \quad n = \overline{2, m-1}$$

$$S_n\left(\tilde{F}_{m-1}(\lambda_\gamma), \lambda_\gamma\right) = \sum_{\gamma_1, \gamma_2, \ldots, \gamma_k} \frac{V_{\gamma - \gamma_1} V_{\gamma_2 - \gamma_1} \ldots V_{\gamma_{n-1} - \gamma_n} V_{\gamma_n - \gamma}}{\prod_{s=1,k} \left(\lambda_\gamma + \tilde{F}_{m-1}(\lambda_\gamma) - \lambda_{\gamma_s}\right)} \quad \text{and} \quad \tilde{F}_n(\lambda_\gamma) = O\left(|\gamma + t|^{-\alpha n}\right).$$

б) For any m-order non-resonance eigenvalue λ_γ there exists an eigenvalue Λ_N of the operator $H_t(a, V(x))$ satisfying to the asymptotics (1.2.40).

Theorem 1.2.3. a) Assume that the eigenvalue $\Lambda_N(t)$ of the operator $H_t(a, V(x))$ satisfies

$$|\Lambda_N - \lambda_\gamma| > 2M, \qquad \left|(\Psi_{N,t}, \varphi_\gamma)\right| > |\gamma + t|^{-\alpha m_0},$$

where $M \geq \sum_{\gamma} |V_{\gamma}| \geq \sup_{h} V(h)$. If λ_{γ} is m-order non-resonance then Λ_N satisfies to the asymptotics

$$\Lambda_N = \lambda_{\gamma} + \tilde{F}_{m-m_0-1}\left(\lambda_{\gamma}\right) + O\left(|\gamma + t|^{-\alpha(m-m_0)}\right). \qquad (1.2.21)$$

б) Eigenvalue $\Lambda_{k\lambda}$, for which is valid

$$\left|\left(\Psi_{k,t}(h), \varphi_{\gamma}(h)\right)\right| = \max_N \left\{\left|\left(\Psi_{k,t}(h), \varphi_{\gamma}(h)\right)\right|, \left|\left(\Psi_{k,t}(h), \varphi^{\gamma}(h)\right)\right|\right\},$$

satisfies to the asymptotics (1.2.41) c $m_0 = \left[\frac{n}{2\alpha}\right]$, where $\left[\frac{n}{2\alpha}\right]$ is an integer part of $\frac{n}{2\alpha}$, $k \equiv k\lambda_{\gamma}$.

Theorem 1.2.4. a) If the eigenvalue λ_{γ} is single resonance i.e. $\gamma + t \in V_{\gamma_1}(\alpha)$, then the corresponding eigenvalue of the operator $H_t(a, V(x))$ satisfies to the asymptotics

$$\Lambda_N = \tilde{\lambda}_{\gamma} + O\left(\left|\tilde{\lambda}_{\gamma}\right|^{\frac{\alpha}{2}}\right), \qquad (1.2.22)$$

where $\tilde{\lambda}_{\gamma}$ is an eigenvalue of the matrix $C = \left(c_{ij}\right)_{i,j=\overline{-p_0, p_0}}$, with $c_{ii} = \left|\lambda_{\gamma + i\gamma_1}\right|$, $c_{ij} = V_{(i-j)\gamma_1}$.

б) For any single resonance eigenvalue λ_{γ} ($\gamma \in V_{\gamma_1}(\alpha)$) there exists an eigenvalue Λ_N of the operator $H_t(a, V(x))$ from $\left[\lambda_{\gamma} - 2M, \lambda_{\gamma} + 2M\right]$ satisfying to the asymptotics (1.2.42).

Now let us assume that $V(x) \in L_2(F)$, i.e. $\sum_{\gamma \in \Gamma} |V_{\gamma}|^2 < \infty$, $\|V(x)\|_{L_2(F)} = M$,

where $$V_{\gamma} = \int_F V(x) e^{-i(\gamma, x)} dx, \quad Q = \{\gamma \in \Gamma : V_{\gamma} \neq 0\}.$$

Denote by $V_{\rho}^{n_0}$ the non-resonance set for the polynomial $V^*(x)$ where

$$V^*(x) = \sum_{\gamma : |\gamma| \leq n_0} V_{\gamma} e^{i(\gamma, x)}, \qquad V_{\rho}^{n_0} = R^2 \setminus \bigcup_{|\gamma| < n_0} N_{\gamma}(\rho),$$

$$N_{\gamma}(\rho) = \left\{x \in R^2 : \left||x|^2 - |x + \gamma|^2\right| < \rho\right\}, \qquad \rho < M_0^2 n_0^4.$$

The following theorem is proved.

Theorem 1.2.5. If $\gamma + t \in V_{\rho}^{n_0}$, then in $\frac{2}{M_0}$ neighborhood of the number λ_{γ} exists at least one eigenvalue $\Lambda_N(t)$ of the operator $H_t(a, V(x))$.

In the third paragraph of the first chapter using the method O.A.Veliev [6] is proved that if $V(x)$ is a trigonometric polynomial or a continuous function, and then the number of gaps in the spectrum of the operator $H(a, V(x))$ is limited. So Bethe-Somerfield conjecture for two-dimensional periodic Pauli operator is true.

In the work [6] using the asymptotic formulas for the eigenvalues of the Schrödinger operator constructively constructed the set B with the following properties:

I. By $\gamma + t \in B$ the eigenvalue $\Lambda_{k(\gamma+t)}$ of the Schrodinger operator is simple and corresponding eigenfunction $\Psi_{k(\gamma+t)}$ is close to the eigenfunction $e^{i(\gamma+t,x)}$ of the Laplace operator

$$\left| \left(\Psi_{k(\gamma+t)}, e^{i(\gamma+t,x)} \right) \right| = 1 + O\left(\left| \gamma + t \right|^{-\alpha} \right), \tag{III}$$

moreover satisfies to the asymptotics

$$\Psi_{k(\gamma+t)} = e^{i(\gamma+t,x)} + \Phi_1(x) + \ldots + \Phi_{m-1}(x) + O\left(\left| \gamma + t \right|^{-m\alpha} \right),$$

where $\Phi_s(x)$ is explicitly calculated and has an order $O\left(\left| \gamma + t \right|^{-s\alpha} \right)$ by $s = 1, 2, \ldots, m-1$.

Thus calculated speed, amperage and impulse corresponding to the state $\Psi_{k(\gamma+t)}$ by $\gamma + t \in B$.

II. The set B constitutes the main part of R^2:

$$\mu\left(B \cap S_\rho \right) = \left(1 + O\left(\rho^{-\alpha} \right) \right) \mu\left(S_\rho \right)$$

And for large ρ contains the intervals

$$A_b(\delta) = \left\{ b + \tau \frac{b}{|b|} : \tau \in [-\delta, \delta] \right\}, \quad (a \in S_\rho = \left\{ x \in R^2 : |x| = \rho \right\}, \delta \approx \rho)$$

From the asymptotics for the eigenvalues of the Schrodinger operator follows that

a) $\Lambda_{k\left(b - \tau \frac{b}{|b|} \right)} < \rho^2 - 2(\rho, a) + a^2 + [\rho, a] \equiv \lambda^\rho$, $\qquad \Lambda_{k\left(b - \tau \frac{b}{|b|} \right)} > \lambda^\rho$,

б) From (III) follows that $\Lambda_{k(\gamma+t)}$ continuously depends on $\gamma + t$ for $\gamma + t \in B$, and so $\gamma + t \in A_b(\delta) \subset B$.

As follows from the considerations above for large ρ there exist eigenvalues $\Lambda_{k(\gamma+t)}$ (where $\gamma + t \in A_b(\delta) \subset B$) coinciding with the number λ^ρ. Thus one can prove that isoenergetic surfaces contain the subset $\{ \gamma + t \in B : \Lambda_{k(\gamma+t)} = \lambda^\rho \}$, the measure of which is asymptotically $(\rho \to \infty)$ close to the measure of the sphere S_ρ. From $\{ \gamma + t \in B : \Lambda_{k(\gamma+t)} = \lambda^\rho \} \neq \emptyset$ for large ρ just follows the validity of the Bethe-Somerfeld conjecture.

Note that the main difficulty here is a construction of the set B with properties I, II. Indeed the property I is equivalent to

$$\sum_{\tilde{\gamma} \in \Gamma / \gamma} \left| \left(\tilde{\Psi}_{k(\gamma+t)}, e^{i(\gamma+t,x)} \right) \right| = O\left(\left| \gamma + t \right|^{-2\alpha} \right), \tag{IV}$$

where $\widetilde{\Psi}_{k(\gamma+t)}$ is arbitrary eigenfunction corresponding to the eigenvalue $\Lambda_{k(\widetilde{\gamma}+t)}$.

The results obtained in the work $[6]$ transferred to two-dimensional periodic Pauli operator in the parallelogram and the following theorems are proved.

Theorem 1.3.3: If $V(x)$ a trigonometric polynomial, then the number of gaps in the spectrum of the operator $H(a,V(x))$ is bounded.

Theorem 1.3.4: If $V(x)$ is a continuous function, then the number of gaps in the spectrum of the operator $H(a,V(x))$ is limited.

Moreover, it is clear that the Theorem 1.3.4 holds for such potentials for which the eigenvalue $\Lambda_\kappa^{V(x)}(t)$ satisfies $\left|\Lambda_\kappa^{V(x)}(t)-\Lambda_\kappa^{V_1(x)}(t)\right|<c$, $c<\delta\rho\approx1$ for some continuous function $V_1(x)$.

Theorem 1.3.5: The set $B=\bigcup_\rho B_\rho(\delta)$, for which the eigenvalue $\Lambda_{\kappa(\gamma+t)}$ is simple and (III) is valid constitutes the main part of R^2. Isoenergetic surface corresponding to the energy λ^ρ contains the set

$$A(\rho)=\left\{\gamma+t\in B:\Lambda_{\kappa(\gamma+t)}=\lambda^\rho\right\},$$

The measure of which for large ρ is close the measure of the sphere S_ρ:

$$\mu(A(\rho))=(1+O(\rho^{-\alpha}))\mu(S_\rho).$$

The second chapter deals with the three-dimensional Pauli operator $H_t(a,V(h))$ generated in $L_2(F)\times L_2(F)$ by the expression (3) and the boundary conditions (4), where $\Omega=\left\{m_1\omega_1+m_2\omega_2+m_3\omega_3:m_1,m_2,m_3\in Z\right\}$, and $\Gamma=\left\{n_1\gamma^1+n_2\gamma^2+n_3\gamma^3:n_1,n_2,n_3\in Z\right\}$ is dual to Ω lattice.

It is easy to find

$$H\left(a,V(h)\right)=\begin{pmatrix}-\Delta+2i(\nabla,a)+a^2+V(h)+[\nabla,a]_z & [\nabla,a]_x-i[\nabla,a]_y \\ [\nabla,a]_x+i[\nabla,a]_y & -\Delta+2i(\nabla,a)+a^2+V(h)-[\nabla,a]_z\end{pmatrix}.$$

In this chapter this operator is studied by two ways. First we represent the Pauli operator $H(a,V(h))$ as follows

$$H\left(a,V(h)\right)=\begin{pmatrix}-\Delta+2i(\nabla,a)+a^2+[\nabla,a]_z & [\nabla,a]_x-i[\nabla,a]_y \\ [\nabla,a]_x+i[\nabla,a]_y & -\Delta+2i(\nabla,a)+a^2-[\nabla,a]_z\end{pmatrix}+\begin{pmatrix}V(h) & 0 \\ 0 & V(h)\end{pmatrix}.$$

In the second scheme we do it as follows

$$H\left(a,V(h)\right)=\begin{pmatrix}-\Delta+2i(\nabla,a)+a^2+[\nabla,a]_z & 0 \\ 0 & -\Delta+2i(\nabla,a)+a^2-[\nabla,a]_z\end{pmatrix}+\begin{pmatrix}V(h) & [\nabla,a]_x-i[\nabla,a]_y \\ [\nabla,a]_x+i[\nabla,a]_y & V(h)\end{pmatrix}.$$

In both schemes the first and second first part of the scheme will be considered as the unperturbed operator and the second part as a perturbation. The first scheme is a simpler form. However, it is applicable only in the case when $a = const$, the second scheme is more general and can be applied when $a = a(h)$.

In the **first paragraph of the second chapter** we obtain asymptotic formulas for the eigenvalues of the order $O(1)$ of three-dimensional operator $H_t(a, V(h))$. First, consider the **first scheme**, i.e. denote by $H_0^t(a)$ the unperturbed operator generated in $L_2(F) \times L_2(F)$ by the expression

$$H_0(a) = (-i\nabla - a)^2 \cdot I + \sigma \cdot B \tag{2.1.6}$$

and boundary conditions (4).

Is possible to calculate the eigenvalues of the operator $H_0^t(a)$ in the form

$$\lambda_{\gamma^-} = |\gamma + t|^2 - 2(\gamma + t, a) + a^2 - i[[\gamma + t, a]], \tag{2.1.10}$$

$$\lambda_{\gamma^+} = |\gamma + t|^2 - 2(\gamma + t, a) + a^2 + i[[\gamma + t, a]]. \tag{2.1.11}$$

It follows immediately that the eigenvalues of the operator $H_t(a, V(h))$ satisfies the following formula

$$\Lambda(\lambda_{\gamma^-}) = |\gamma + t|^2 - 2(\gamma + t, a) + a^2 - i[[\gamma + t, a]] + O(1), \tag{2.1.12}$$

$$\Lambda(\lambda_{\gamma^+}) = |\gamma + t|^2 - 2(\gamma + t, a) + a^2 + i[[\gamma + t, a]] + O(1). \tag{2.1.13}$$

In the **second scheme** we denote by $M_t(a)$ the operator generated in $L_2(F) \times L_2(F)$ by the expression

$$M(a) = \begin{pmatrix} -\Delta + 2i(\nabla, a) + a^2 + [\nabla, a]_z & 0 \\ 0 & -\Delta + 2i(\nabla, a) + a^2 - [\nabla, a]_z \end{pmatrix} \tag{2.1.15}$$

And boundary conditions (4).

Denote by $N(h)$ the expression

$$\begin{pmatrix} V(h) & [\nabla, a]_x - i[\nabla, a]_y \\ [\nabla, a]_x + i[\nabla, a]_y & V(h) \end{pmatrix}. \tag{2.1.16}$$

By the elementary computations we find that the eigenvalues and eigenvectors of the operator $M_t(a)$ are as follows

$$\phi_\gamma(h) = \begin{pmatrix} 0 \\ e^{i(\gamma + t, h)} \end{pmatrix}, \quad \text{where } \gamma \in \Gamma, \ t \in F, \tag{2.1.17}$$

$$\lambda_\gamma = |\gamma + t|^2 - 2(\gamma + t, a) + a^2 - i[\gamma + t, a]_z, \tag{2.1.18}$$

$$\phi^\gamma(h) = \begin{pmatrix} e^{i(\gamma + t, h)} \\ 0 \end{pmatrix}, \quad \text{where } \gamma \in \Gamma, \ t \in F, \tag{2.1.19}$$

$$\lambda^\gamma = |\gamma + t|^2 - 2(\gamma + t, a) + a^2 + i[\gamma + t, a]_z, \tag{2.1.20}$$

As in the two-dimensional case, we assume that the potential $V(h)$ is a trigonometric polynomial.

We use the following notations

$$c^\gamma = [\gamma + t, a]_x - i[\gamma + t, a]_y \tag{2.1.22}$$

$$b^\gamma = [\gamma + t, a]_x + i[\gamma + t, a]_y \tag{2.1.23}$$

It proved that if $\left|[\gamma + t, a]_z\right| > \left|[\gamma + t, a]_x\right| + \left|[\gamma + t, a]_y\right|$, i.e. $\left|\lambda^\gamma - \lambda_\gamma\right| > 2c^\gamma$ and $\gamma + t \notin V^m(\alpha)$,

where $V^m(\alpha) = \bigcup_{\gamma_1 \in Q^m}(V_{\gamma_1}(\alpha))$, $V_{\gamma_1}(\alpha) = \left\{h \in R^3 : \left||h|^2 - |h + \gamma_1|^2\right| < |x|^\alpha\right\}$, then for any eigenvalue $\lambda_\gamma(\lambda^\gamma)$

of the operator $M_t(a)$ corresponding to the eigenvalue $\Lambda_N(\lambda_\gamma)$ $(\Lambda_N(\lambda^\gamma))$ of the operator

$H_t(a, V(h))$ satisfies the formula

$$\Lambda_N(\lambda_\gamma) = \lambda_\gamma + [\gamma + t, a]_z \mp \sqrt{[\gamma + t, a]_z^2 + c^\gamma b^\gamma} + O(1). \tag{2.1.31}$$

Since $c^\gamma b^\gamma = [\gamma + t, a]_x^2 + [\gamma + t, a]_y^2$, then (2.1.31) may be written in the form

$$\Lambda_N(\lambda_\gamma) = \lambda_\gamma + [\gamma + t, a]_z \mp \left|[\gamma + t, a]\right| + O(1). \tag{2.1.32}$$

For any eigenvalue λ_γ of the operator $M_t(a)$ satisfying the above conditions, there exists eigenvalue Λ_N of the operator $H_t(a, V(h))$ satisfying the asymptotic formula (2.1.31) or (2.1.32).

In the **second paragraph of the second chapter**, again using the above scheme, we obtain the asymptotic formula of high order for some series of eigenvalues of the three-dimensional operator $H_t(a, V(h))$.

Using (2.1.10) and (2.1.11), we find the eigenfunctions of the operator $H_0^t(a)$

$$\varphi_+(h) = \begin{pmatrix} \dfrac{i[\gamma + t, a]_x - [\gamma + t, a]_y}{A} e^{i(\gamma + t, h)} \\ \dfrac{i[\gamma + t, a]_z - \left|[\gamma + t, a]\right|}{A} e^{i(\gamma + t, h)} \end{pmatrix},$$

$$\varphi_-(h)=\begin{pmatrix} -\dfrac{i[\gamma+t,a]_x-[\gamma+t,a]_y}{A}e^{i(\gamma+t,h)} \\ -\dfrac{i[\gamma+t,a]_z-[[\gamma+t,a]]}{A}e^{i(\gamma+t,h)} \end{pmatrix},$$

where

$$A=\sqrt{2[\gamma+t,a]_y^2-2i[\gamma+t,a]_z[\gamma+t,a]-2i[\gamma+t,a]_x[\gamma+t,a]_y}.$$

Let

$$V_\gamma(\alpha)=\left\{h\in R^3:\left|\|h\|^2-|h+\gamma|^2\right|<|x|^\alpha\right\}.$$

Denote

$$V^m(\alpha)=\bigcup_{\gamma_1\in Q^m}(V_{\gamma_1}(\alpha)),\qquad \overline{V}^m(\alpha)=R^3\setminus V^m(\alpha).$$

Assume that

$$\|[\gamma+t,a]\|<\frac{1}{5}|\gamma+t|^\alpha,\text{ i.e. }\left|\lambda_{\gamma^+}-\lambda_{\gamma^-}\right|<\frac{2}{5}|\gamma+t|^\alpha,\tag{2.2.4}$$

or

$$\sin\beta<c|\gamma+t|^{\alpha-1},\qquad\text{where}\quad\beta=(\overrightarrow{\gamma+t},\ \vec{a}).$$

If $\gamma+t\in V^m(\alpha)$ and the condition (2.2.4) is valid then as defined the eigenvalues λ_{γ^+} and λ_{γ^-} of the operator $H_0^t(a)$ are m-order non-resonance.

The following theorem is proved (analogues of the Theorem 1.2.2).

Theorem 2.2.1. a) If the eigenvalue λ_{γ^+} (λ_{γ^-}) is m-order non-resonance (i.e. $\gamma+t\notin V^m(\alpha)$) and the condition (2.2.4) holds true, the eigenvalue $\Lambda(\lambda_{\gamma^+})$ ($\Lambda(\lambda_{\gamma^-})$) of the operator $H_t(a,V(h))$ satisfies the asymptotics

$$\Lambda(\lambda_{\gamma^+})=\lambda_{\gamma^+}+\tilde{F}_{m-1}(\lambda_{\gamma^+})+O(|\gamma+t|^{-\alpha m}),\tag{2.2.5}$$

where $\tilde{F}_k$ is explicitly expressed by λ_{γ^+}, λ_{γ^-} and $V(h)$, and has order $O(|\gamma+t|^{-\alpha k})$.

б) For any m-order non-resonance eigenvalue λ_γ exists eigenvalue Λ_N of the operator $H_t(a,V(h))$ satisfying the asymptotics (2.2.5).

In the first section of the second chapter we have shown that the eigenvalues and eigenfunctions of the operator $M_t(\alpha)$ satisfy the formula (2.1.17) - (2.1.20).

To obtain higher-order asymptotic formulas in the scheme will impose the following conditions:

I. $a \in XOY$, т.е. $a = (a_1, a_2, 0)$;

II. $v = (0, 0, 1) \notin Q^{(m)}$;

III. $[\gamma + t, a]_z < c \cdot |\gamma + t|^{\alpha^1}$, т.е. $|\lambda_\gamma - \lambda^\gamma| < c|\gamma + t|^{\alpha'}$.

Here III is an analogue of the condition (2.2.4).

The following theorems have been proved.

Theorem 2.2.3. If the conditions I – III hold, $\gamma + t \in V_v(\alpha)$ and $|\gamma + t|$ is large enough then

the corresponding eigenvalue $\Lambda_N(\lambda_\gamma)$ of the operator $H_t(a, V(h))$ satisfies to the formula

$$\Lambda_N(\lambda_\gamma) = \lambda_\gamma + \sum_{n=1}^{m-1} \overline{F}_n(\lambda_\gamma, \lambda^\gamma, V(x)) + O\left(|\gamma + t|^{-(\alpha - \alpha')m}\right), \tag{2.2.43}$$

where $\overline{F}_n(\lambda_\gamma, \lambda^\gamma, V(h))$ is explicitly expressed by λ_γ, λ^γ, $V(h)$ and has order $O\left(|\gamma + t|^{-(\alpha - \alpha')n}\right)$.

Theorem 2.2.4. EIf the conditions of Theorem 2.2.3. are satisfied then for any λ_γ there

exists an eigenvalue $\Lambda_N(\lambda_\gamma)$ of the operator $H_t(a, V(h))$ that satisfies to the formula (2.2.43).

Remark 2.1. By the first and second scheme it is possible to show if $\Lambda_N(\lambda_\gamma)$ is such that

$$\left|\left(\Psi_N(h), \varphi_\gamma(h)\right)\right| > |\gamma + t|^{-s(\alpha - \alpha')},$$

Takes place then $\Lambda_N(\lambda_\gamma)$ satisfies to the asymptotic formula (2.2.83) obtained by replacing

$O\left(|\gamma + t|^{-(\alpha - \alpha')m}\right)$ by $O\left(|\gamma + t|^{(s-n)(\alpha - \alpha')}\right)$.

The similar formula is true also for $\Lambda_N(\lambda^\gamma)$.

In the third paragraph of the second part the obtained results are generalized for the

operators type of $H_t^l(a, V(h))$ generated in $L_2(F) \times L_2(F)$ by the expression

$$H^l(a, V(h)) = \left((-i\nabla - a)^{2l} + V(h)\right) \cdot I + \sigma \cdot B, \quad l > 1$$

and boundary conditions (4).

Let us denote the non-perturbed operator by $M_t^l(a)$ where

$$M^l(a) = (-i\nabla - a)^{2l} \cdot I + [\nabla, a]_z \cdot \sigma_3.$$

Eigenvalues of this operator indeed are

$$\lambda_\gamma = \left(-i|\gamma + t| - a\right)^{2l} - i[\gamma + t, a]_z, \tag{2.3.9}$$

$$\lambda^\gamma = \left(-i|\gamma + t| - a\right)^{2l} + i[\gamma + t, a]_z; \tag{2.3.10}$$

$$\phi_\gamma(h) = \begin{pmatrix} 0 \\ e^{i(\gamma+t,h)} \end{pmatrix}, \quad \text{где} \quad \gamma \in \Gamma, \quad t \in F, \tag{2.3.11}$$

$$\phi^\gamma(h) = \begin{pmatrix} e^{i(\gamma+t,h)} \\ 0 \end{pmatrix}, \quad \text{где} \quad \gamma \in \Gamma, \quad t \in F. \tag{2.3.12}$$

Denote:

$$V_\gamma(\alpha) = \left\{ h \in R^3 : \left| |h|^{2l} - |h+\gamma|^{2l} \right| < |h|^{\alpha l} \right\}, \quad \alpha l > 1,$$

$$V^m(\alpha) = \bigcup_{\gamma \in Q^m} V_\gamma(\alpha), \quad \overline{\overline{V^m(\alpha)}} = R^3 \setminus V^m(\alpha).$$

It is proved the following

Theorem 2.8. If $\gamma + t \in \overline{\overline{V^m(\alpha)}}$ i.e. λ_γ is non-resonance eigenvalue of the operator $M_t^l(a)$, then corresponding eigenvalue $\Lambda(\lambda_\gamma)$ of the operator $H_t^l(a,V(h))$ (i.e. Pauli type operator) satisfies to the asymptotic formula

$$\Lambda(\lambda_\gamma) = \lambda_\gamma + \sum_{k=1}^{m-1} \overline{F_n}(\lambda_\gamma, \lambda^\gamma, V(h)) + O\left(|\gamma+t|^{m(1-\alpha l)}\right),$$

where $\overline{F_n}(\lambda_\gamma, \lambda^\gamma, V(h))$ is explicitly expressed through λ_γ, λ^γ, $V(h)$ and has an order $O\left(|\gamma+t|^{-\alpha l m+n}\right)$.

Remark. The similar result is valid also for the eigenvalue $\Lambda_N(\lambda^\gamma)$.

The following denotations have been used in the work.

By $c_1, c_2, \dots$ are denoted constants which have different values in different parts of the book. The definition $f(\zeta) \square g(\zeta)$ means that there exist positive constants c_1, c_2 such that $c_1|g(\zeta)| < |f(\zeta)| < c_2|g(\zeta)|$.

The eigenvalues under investigation $\Lambda_N, \lambda_\gamma, \lambda^\gamma, \lambda_{\gamma^*}, \lambda_{\gamma^-}, |\gamma+t|^2, \dots$ and spectral $\rho, |\gamma+t|$ are always assumed large enough. The definition $f(\zeta) \square g(\zeta)$ means that $f(\zeta) = o(g(\zeta))$, $\zeta \to \infty$.

$\mu_k(A)$ stands for k-dimensional Lebequie measure of the set A. If from the text is clear that which measure is assumed the index k is omitted.

PART 1

On a spectrum of the periodic Pauli operator in parallelogram
(Bethe-Sommerfeld hypothesis)

In this part we consider the Pauli operator $H_t(a, V(x))$ generated in $L_2(F) \times L_2(F)$ by the expression

$$H(a, V(x)) = \left((-i\nabla - a)^2 + V(x) \right) \cdot I + \sigma \cdot B \tag{1.1.1}$$

and boundary conditions

$$u(x + \omega_j) = e^{2\pi i t_j} u(x), \tag{1.1.2}$$

where $x \in R^2$, $B = [grad \times a]$ is a magnet field generated by the vector potential $a = (a_1, a_2)$,

$$\sigma = \begin{pmatrix} 1 & 0 \\ 0 & -1 \end{pmatrix}, \qquad I = \begin{pmatrix} 1 & 0 \\ 0 & 1 \end{pmatrix},$$

F is a fundamental domain of some lattice $\Omega = \{ m_1\omega_1 + m_2\omega_2 : m_1, m_2 \in Z \}$ i.e. parallelogram, $t = t_1\gamma^1 + t_2\gamma^2$ a γ^1, γ^2 are dual vectors to ω_1, ω_2, F^* is a fundamental domain for the lattice $\Gamma = \{ n_1\gamma^1 + n_2\gamma^2 : n_1, n_2 \in Z \}$, $V(x)$ -periodic and smooth enough function.

First, in the first section, we show that the spectrum of the operator $H(a, V(x))$ generated in $L_2(F) \times L_2(F)$ by the expression (1.1) is the union of the spectra of the operators $H_t(a, V(x))$.

In the second section we obtain asymptotic formulas for the eigenvalues of the Pauli operator, i.e. operator $H_t(a, V(x))$ in the parallelogram when $V(x)$ is trigonometric polynomial an $V(x) \in L_2(F)$. And in the third section, we show that the number of gaps in the spectrum of two-dimensional Pauli operator in the parallelogram is limited i.e. the Bethe-Sommerfeld conjecture is true.

1.1. Decomposition of the periodic Pauli operator in layers

Let us prove that the spectrum of the operator $H(a, V(x))$ generated in $L_2(F) \times L_2(F)$ by the expression (1) is a union of the spectrums of $H_t(a, V(x))$.

First we give some definitions from $[26]$.

Definition. Let be connected domain in the complex plane and let each $\beta \in R$ is given a closed operator $T(\beta)$ with nonempty resolvent set. We say that $T(\beta)$ is analytic family of type (A), if and only if

    1) The operator definition domain of $T(\beta)$ is some independent on β set;

    2) $T(\beta)\Psi$ is a vector valued function of β for each $\Psi \in D$.

Definition. Operator-valued function $T(\beta)$ (possibly unbounded) in the complex plane R is called an analytic family or analytic family in the sense of Kato if and only if

    1) for each $\beta \in R$ the operator $T(\beta)$ is closed and its resolvent set is not empty;

    2) for each $\beta_0 \in R$ there exists some $\lambda_0 \in \rho(T(\beta_0))$ such that $\lambda_0 \in \rho(T(\beta))$ by β close to β_0, and $(T(\beta) - \lambda_0)^{-1}$ is an analytic operator valued function of β in the neubourhood of β_0.

It is clear that each (A) family is a Kato family.

Let H' be separable Hilbert space, $<M, \mu>$ - space with σ-bounded measure, $L^2(M, d\mu; H')$ - Hilbert space of quadratic integrable H' valued functons. Note that if μ is a sum of pointwise measures consolidated in the pints $m_1, m_2, ..., m_n$ then any $f \in L^2(M, d\mu; H')$ defined by the collection $<f(m_1), f(m_2), ..., f(m_n)>$ is isomorph to the direct sum $\bigoplus_{i=1}^{n}(H'_i) = H'$. Then $L^2(M, d\mu; H')$ for more general measures μ is some kind of "continuous direct sum" but with same terms. That is why we call $H \equiv L^2(M, d\mu; H')$ direct integral of the spaces with same layers and write $H = \int_{M}^{\oplus} H' d\mu$.

This is done in order to shift the focus from the points of the space M on the "layer" H'. We are interested in a special class of operators on H. The function $A(\cdot)$ from M to $\mathfrak{I}(H')$ is called a measurable if and only if $(\phi, A(\cdot)\varphi)$ is a measurable function for any $\varphi, \phi \in H'$. The symbol $L^{\infty}(M, d\mu; \mathfrak{I}(H'))$ denotes the space of (equivalence class is defined as equal almost everywhere) measurable functions from M to $\mathfrak{I}(H')$ such that

$$\|A\|_{\infty} = ess \sup \|A(m)\|_{\mathfrak{I}(H')} < \infty .$$

Definition. We say that the bounded operator A on $H = \int_{M}^{\oplus} H' d\mu$ is decomposed by the direct integral of the spaces if and only if there exists a function $A(\cdot) \in L^{\infty}\left(M, d\mu; \mathfrak{I}(H')\right)$ such that

$$\left(A\varphi\right)(m) = A(m)\varphi(m). \tag{1.1.3}$$

for any $\phi \in H$

In this case we call the function A decomposable and write $A = \int_{M}^{\oplus} A(m) d\mu(m),$ moreover $A(m)$ are called the layers of the operator A.

Note that each function $A(\cdot)$ from $L^{\infty}\left(M, d\mu; \mathfrak{I}(H')\right)$ is associated with some decomposable operator.

Theorem. If $A(\cdot) \in L^{\infty}\left(M, d\mu; \mathfrak{I}(H')\right)$, then there exists single defined decomposable operator $A \in \mathfrak{I}(H)$, such that (3) holds true. Moreover

$$\left\|A(m)\right\|_{\mathfrak{I}(H')} = \left\|A\right\|_{\infty}. \tag{1.1.4}$$

This theorem establishes an isometrical isomorphism between $L^{\infty}\left(M, d\mu; \mathfrak{I}(H')\right)$ and the set of decomposable operators on $\int_{M}^{\oplus} H' d\mu$. Both these spaces indeed are algebras and it is easy to see that in the described isomorphism the algebraic structures are valid. $L^{\infty}\left(M, d\mu; C\right)$ is a natural subalgebra in $L^{\infty}\left(M, d\mu; \mathfrak{I}(H')\right)$, corresponding to the decomposable operators all layers of which are multiple to the identity operator.

Theorem. Let $H = \int_{M}^{\oplus} H' d\mu$ where $< M, \mu >$ is a separable space with bounded measure σ and H' is separable. Let $\aleph$ be algebra of decomposable operators with layers that are multiples of the identity operator. In this case, $A \in \mathfrak{I}(H)$ is decomposable if and only if when A commutes with every operator from $\aleph$.

Used by us following construction is essentially based on the fact that $U(t)$ generates the algebra isomorphic to the algebra $\aleph$ corresponding to the suitable decomposition $H = L^2\left(R^2, dx\right)$ into a direct integral with the same layers.

Definition. The function $A(\cdot)$ from the space with measure $<M, \mu>$ in the set of self-adjoint (not necessary bounded) operators on the Hilbert space H' is measurable if and only if the function $\left(A(\cdot)+I\right)^{-1}$ is measurable.

Knowing such a function, we define the operator A in $H = \int_M^\oplus H' d\mu$ with a domain

$$D(A) = \left\{ \phi \in H \mid \varphi(m) \in D\left(A(m)\right) n..s., \int_M \left\| A(m)\varphi(m) \right\|^2_{H'} d\mu(m) < \infty \right\}, \tag{1.1.5}$$

By the relation $(A\varphi)(m) = A(m)\varphi(m)$ and write

$$A = \int_M^\oplus A(m) d\mu(m). \tag{1.1.6}$$

For such defined operators is proved

Theorem 1.1.1. Let $A = \int_M^\oplus A(m) d\mu(m)$, where $A(\cdot)$ is measurable and $A(m)$ is self-adjoint operator for each m. Then $\lambda \in \sigma(A)$ if and only if

$$\mu\left(\left\{ m \in M \mid \sigma\left(A(m)\right) \cap (\lambda - \varepsilon, \lambda + \varepsilon) \neq \emptyset \right\}\right) > 0. \tag{1.1.7}$$

for any $\varepsilon > 0$

Now, these results we carry over to the Pauli operator with a periodic potential $V(x)$, i.e. it is assumed that potential satisfies

$$V(x + \varpi_i) = V(x) \; i = 1,2. \tag{1.1.8}$$

for some basis $\{\varpi_1\}, \{\varpi_2\} \in R^2$.

Thus, the spectral properties of the Pauli operator is very sensitive to the behavior of $V(x)$ at the infinity, and as $V(x)$ satisfying (1.1.8) has no limit as in any direction by $x \to \infty$, it can be expected that the analysis of periodic Pauli operators will be difficult enough.

The property, which allows one to analyze the operator $H\left(a, V(x)\right)$ when $V(x)$ is a periodic function, is that it is symmetrical with respect to the large group actions. In fact, considering $(U(t)\varphi)(x) = \varphi(x + t_1 a_1 + t_2 a_2)$ where $t \in Z^2$ we see that (formally)

$$U(t) H\left(a, V(x)\right) = H\left(a, V(x)\right) U(t). \tag{1.1.9}$$

In order to use the above theory it is that $u(x)$ be periodic. So let us make the replacement $u(x) = e^{i<t,x>} u_1(x)$.

From this we obtain that

$$u_1(x) = e^{-i<t,x>}u(x).$$ (1.1.10)

Now using the formulas (1.1.2) and (1.1.10) we get

$$u_1(x+\varpi_j) = e^{-i<t,x>}u(x+\varpi_j) = e^{-i<t,x>}e^{-i<t,\varpi_j>}u(x+\varpi_j) =$$

$$= e^{-i<t,x>}e^{-2\pi i t_j}u(x) = e^{-i<t,x>}u(x) = u_1(x), \quad j=1,2.$$

This proves that $u_1(x)$ is periodic with respect to $\{\varpi_1\}, \{\varpi_2\}$.

From (10) is obtained

$$\frac{\partial u(x)}{\partial x_1} = it_1 e^{i<t,x>}u_1(x) + \frac{\partial u_1(x)}{\partial x_1}e^{i<t,x>}, \quad \frac{\partial u(x)}{\partial x_2} = it_2 e^{i<t,x>}u_1(x) + \frac{\partial u_1(x)}{\partial x_2}e^{i<t,x>},$$

$$\frac{\partial^2 u(x)}{\partial x_1^2} = -t_1^2 e^{i<t,x>}u_1(x) + 2it_1 e^{i<t,x>}\frac{\partial u_1(x)}{\partial x_1} + e^{i<t,x>}\frac{\partial^2 u_1(x)}{\partial x_1^2},$$

$$\frac{\partial^2 u(x)}{\partial x_2^2} = -t_2^2 e^{i<t,x>}u_1(x) + 2it_2 e^{i<t,x>}\frac{\partial u_1(x)}{\partial x_2} + e^{i<t,x>}\frac{\partial^2 u_1(x)}{\partial x_2^2}.$$

Consideration this in (1.1.1) gives

$$H(a,V(x))u_1(x) = -(t_1^2+t_2^2)u_1(x) + 2i\left(t_1\frac{\partial u_1(x)}{\partial x_1} + t_2\frac{\partial u_1(x)}{\partial x_2}\right) + \frac{\partial^2 u_1(x)}{\partial x_1^2} + \frac{\partial^2 u_1(x)}{\partial x_2^2} +$$

$$+(2ia_1 \pm a_2)\left(it_1 u_1(x) + \frac{\partial u_1(x)}{\partial x_1}\right) + (2ia_2 \mp a_1)\left(it_2 u_1(x) + \frac{\partial u_1(x)}{\partial x_2}\right) + a^2 u_1(x).$$ (1.1.11)

We define this operator by $H(t)$, where $t=(t_1,t_2)$. Now we prove that $H(t)$ is an analytic family type of (A).

1) The operator definition domain of $H(t)$ does not depend on t, therefore F does not depend on t.

2) Now let us prove that $H(t)\varphi$ is vector valued analytic function of t for each $\varphi \in L_2(F) \times L_2(F)$.

From the formula (1.1.11) follows that $H(t)\varphi$ is analytic function relative to t_1 when t_2 is fixed and vice versa.

Let ϖ_1, ϖ_2 be some basis in R^2, and γ^1, γ^2 be an orthogonal to ϖ_1, ϖ_2 basis i.e. $(\varpi_j, \gamma^i) = 2\pi\delta_{ij}$, $V(x)$ - a function on R^2 such that $V(x+\varpi_j) = V(x)$, $H' = L_2(Z^2) \times L_2(Z^2)$, $H = \int_{F^*}^{\oplus} H' d^2\kappa$ and V_γ - Fourier coefficients of $V(x)$ as a function on F, i.e. for any $m \in Z^2$

$$V_\gamma = \int_F e^{-i<\gamma,x>}V(x)dx.$$ (1.1.12)

For $\kappa \in F^*$ define the operator $H(\kappa)$ on H' as

$$
(H(\kappa)g)_m = \begin{pmatrix} \left(\kappa + \sum_{i=1}^{2}\gamma^i\right)^2 g_m + (2ia_1 + a_2)\gamma^1 g_m + (2ia_2 - a_1)\gamma^2 g_m + a^2 g_m + \sum_{i=1}^{2} V_\gamma g_{m-\gamma} \\ \left(\kappa + \sum_{i=1}^{2}\gamma^i\right)^2 g_m + (2ia_1 - a_2)\gamma^1 g_m + (2ia_2 + a_1)\gamma^2 g_m + a^2 g_m + \sum_{i=1}^{2} V_\gamma g_{m-\gamma} \end{pmatrix},
$$
(1.1.13)

with definition domain $D_0 = \left\{ g \in H' : \sum m^2 |g_m|^2 < \infty \right\}$.

Finally suppose that $U : L_2\left(R^2\right) \to H$ is given by relation

$$
[(Uf)(\kappa)]_m = \hat{f}\left(\kappa + m_1 \gamma^1 + m_2 \gamma^2\right)\cdot I, \text{ where } I = \begin{pmatrix} 1 & 0 \\ 0 & 1 \end{pmatrix}.
$$

Now we prove that U is an unitary operator and

$$
U \cdot \begin{pmatrix} -\Delta + 2i(\nabla,a) + a^2 - [\nabla,a] + V(x) & 0 \\ 0 & -\Delta + 2i(\nabla,a) + a^2 + [\nabla,a] + V(x) \end{pmatrix} \cdot U^{-1} = \int_{F^*}^{\oplus} H(\kappa)d^2\kappa.
$$
(1.1.14)

Unitarity of U follows from Plansherel's theorem moreover it is clear that

$$
\left[\left(U\cdot(-\Delta)\cdot U^{-1}\right)g\right](\kappa)_m = \begin{pmatrix} \left(\kappa + m_1\gamma^1 + m_2\gamma^2\right)^2 g(\kappa)_m \\ \left(\kappa + m_1\gamma^1 + m_2\gamma^2\right)^2 g(\kappa)_m \end{pmatrix},
$$
(1.1.15)

since $-\hat{\Delta}f(l) = l^2 \hat{f}(l)$. Considering this we get

$$
\left[\left(U\cdot\left(2i(\nabla,a) + a^2 \mp [\nabla,a]\right)\cdot U^{-1}\right)g\right](\kappa)_m =
$$

$$
= \begin{pmatrix} (2ia_1 + a_2)\gamma^1 g(\kappa)_m + (2ia_2 - a_1)\gamma^2 g(\kappa)_m + a^2 g(\kappa)_m \\ (2ia_1 - a_2)\gamma^1 g(\kappa)_m + (2ia_2 + a_1)\gamma^2 g(\kappa)_m + a^2 g(\kappa)_m \end{pmatrix}.
$$
(1.1.16)

Let us prove that

$$
\left[\left(U\cdot V(x)U^{-1}\right)g\right](\kappa)_m = \begin{pmatrix} \sum_{\gamma \in Z^2} V_\gamma g_{m-\gamma}(\kappa) \\ \sum_{\gamma \in Z^2} V_\gamma g_{m-\gamma}(\kappa) \end{pmatrix}.
$$
(1.1.17)

For this purpose it is enough to show that for $f \in Z\left(R^2\right)$

$$
\hat{V}f(\kappa) = \sum_{\alpha \in Z^2} V_\gamma f\left(\kappa - \alpha_1\gamma^1 - \alpha_2\gamma^2\right).
$$
(1.1.18)

To finish the proof of (1.1.18) only need to ensure that the generalized function with moderate growth has Fourier transformation

$$
\hat{V}(\kappa) = 2\pi \cdot \sum_{\alpha \in Z^2} V_\gamma \delta\left(\kappa - \alpha_1\gamma^1 - \alpha_2\gamma^2\right).
$$

But this is true as well as in one dimensional case the Fourier series

$$V(x) = \sum_{\alpha \in Z^2} V_\gamma e^{i<\alpha\gamma,x>}$$

converges locally сходится in the sense of L_2, since $V(x)$ belongs to L_2 locally uniformly.

From (1.1.15), (1.1.16), (1.1.17) follows that (1.1.14) is true. Additionally we know that

$$L_2\left(R^2\right) \times L_2\left(R^2\right) = \int_\Omega^\oplus L_2\left(F\right) \times L_2\left(F\right) d^2\mu. \tag{1.1.19}$$

As we can see from (1.1.14) и (1.1.19) Theorem 1 is valid i.e. $\lambda \in \sigma\left(H(t)\right)$ if and only if when for each

$$\mu\left(\left\{t \in F \,\left|\, \sigma\left(H(t)\right) \cap \left(\lambda - \varepsilon, \lambda + \varepsilon\right) \neq \emptyset\right.\right\}\right) > 0. \tag{1.1.20}$$

Now using (A) type analyticity of the operator $H(t) = H(a, V(x))$ we prove that from (1.1.20) follows

Theorem 1.1.2.

$$\sigma\left(H\left(a, V(x)\right)\right) = \bigcup_{t \in F} \sigma\left(H_t\left(a, V(x)\right)\right). \tag{1.1.21}$$

Proof. a)First assume that $\lambda \in \bigcup_{t \in F} \sigma\left(H_t\left(a, V(x)\right)\right)$. From this we get the existence of t_0 for

which $\lambda \in \sigma\left(H_{t_0}\left(a, V(x)\right)\right)$.

Now let $\lambda = \lambda_n\left(t_0\right)$. It is clear that $\lambda_n\left(t_0\right)$ is not constant. не постоянная. In $[1]$ is proved

that $\left|\dfrac{\partial \lambda_n(t)}{\partial t}\right| < c\left(\left|\lambda_n(t)\right| + |t| + 1\right)$.

Denote $M \equiv \left|\lambda_n(t)\right| + |t| + 1$. Then for $\tau \in \left[-\delta, \delta\right]$, where $\delta < \dfrac{\varepsilon}{M}$, takes place

$\lambda\left(t_0 + \tau\right)\tau \in \left[\lambda_0 - \varepsilon, \lambda_0 + \varepsilon\right]$, i.e.

$$\mu\left(\left\{t \in F \,\left|\, \sigma\left(H_t\right) \cap \left(\lambda - \varepsilon, \lambda + \varepsilon\right) \neq \emptyset\right.\right\}\right) > 2\delta > 0.$$

From this considering Theorem 1.1.1 we obtain that $\lambda \in \sigma\left(H\left(a, V(x)\right)\right)$.

б) Suppose that $\lambda \in \sigma\left(H\left(a, V(x)\right)\right)$. Then as follows from Theorem 1.1.1

$\forall \varepsilon_\kappa > 0 \left(\lim\limits_{k \to \infty} \varepsilon_\kappa = 0\right)$, $\mu\left(\left\{t \in F \,\left|\, \sigma\left(H_t\right) \cap \left(\lambda - \varepsilon_\kappa, \lambda + \varepsilon_\kappa\right) \neq \emptyset\right.\right\}\right) > 0$, i.e. there exists t_κ

and N_κ, such that

$$\lambda_{N_\kappa}\left(t_\kappa\right) \in \sigma\left(H_{t_0}\left(a, V(x)\right)\right) \Rightarrow \lambda_{N_\kappa}\left(t_\kappa\right) \in \left(\lambda - \varepsilon_\kappa, \lambda + \varepsilon_\kappa\right).$$

Since such indexes N_κ are more than finite number there exists a number N with corresponding $\{t_{n_\kappa}\}$ from $\{t_k\}$, i.e.. $\lambda_N(t_{n_\kappa}) \in (\lambda - \varepsilon_\kappa, \lambda + \varepsilon_\kappa)$.

From the sequence $\{t_{n_\kappa}\}$ one can choose a subsequence that converges to t_0 (that subsequence we also denote by t_{n_κ} i.e. $\lim_{k \to \infty} t_{n_\kappa} = t_0$). Then considering the continuity of $\lambda_N(t)$ we obtain that

$$\lim_{k \to \infty} \lambda_N(t_{n_\kappa}) = \lambda_N(t_0), \text{ i.e. } \lambda_0 \in \sigma\big(H_{t_0}(a, V(x))\big) \Rightarrow \lambda_0 \in \bigcup_{t \in F} \sigma\big(H_t(a, V(x))\big).$$

This proves the theorem.

Thus we proved that the spectrum of the operator $H(a, V(x))$ generated in $L_2(F) \times L_2(F)$ by the expression (1) indeed is a union of the spectrums of $H_t(a, V(x))$.

So in future we'll study only the spectrum of the operator $H_t(a, V(x))$.

1.2. Asymptotic formulas for the eigenvalues of the two-dimensional Pauli operator in the parallelogram

In this chapter we consider operator $H_t(a, V(x))$ generated in $L_2(F) \times L_2(F)$ by the expression

$$H(a, V(x)) = \big((-i\nabla - a)^2 + V(x)\big) \cdot I + \sigma \cdot B \tag{1.1.1}$$

and boundary conditions

$$u(x + \omega_j) = e^{2\pi i t_j} u(x) \tag{1.1.2}$$

where $x \in R^2, B = [grad \times a]$ is a magnet field generated by the vector potential $a = (a_1, a_2)$,

$$\sigma = \begin{pmatrix} 1 & 0 \\ 0 & -1 \end{pmatrix}, \qquad I = \begin{pmatrix} 1 & 0 \\ 0 & 1 \end{pmatrix},$$

F is a fundamental domain for some lattice $\Omega = \{m_1\omega_1 + m_2\omega_2 : m_1, m_2 \in Z\}$, i.e. parallelogram. $t = t_1\gamma^1 + t_2\gamma^2$ a γ^1, γ^2 are biorthogonal to ω_1, ω_2 vectors, F^* - fundamental domain of the lattice $\Gamma = \{n_1\gamma^1 + n_2\gamma^2 : n_1, n_2 \in Z\}$, $V(x)$ is periodic, smooth enough function. Without loss of generality, we assume that $\mu(F) = 1$.

Let us denote by $H_0'(a)$ the operator generated in $L_2(F) \times L_2(F)$ by the expression

$$H_0(a) = (-i\nabla - a)^2 \cdot I + \sigma \cdot B \tag{1.2.2}$$

and boundary conditions (1.1.2).

By the elementary calculation we find that

$$H_0(a) = \begin{pmatrix} (-i\nabla - a)^2 + \dfrac{\partial}{\partial x_1} a_2 - \dfrac{\partial}{\partial x_2} a_1 & 0 \\ 0 & (-i\nabla - a)^2 - \dfrac{\partial}{\partial x_1} a_2 + \dfrac{\partial}{\partial x_2} a_1 \end{pmatrix}. \qquad (1.2.3)$$

Considering

$$H_0^t(a)\begin{pmatrix} u_1(x) \\ u_2(x) \end{pmatrix} = \lambda \begin{pmatrix} u_1(x) \\ u_2(x) \end{pmatrix}. \qquad (1.2.4)$$

from (1.2.3) and (1.2.4) we get

$$\begin{cases} \left(-\Delta + 2i(\nabla, a) + a^2 + \dfrac{\partial}{\partial x_1} a_2 - \dfrac{\partial}{\partial x_2} a_1 \right) u_1(x) = \lambda u_1(x) \\ \left(-\Delta + 2i(\nabla, a) + a^2 - \dfrac{\partial}{\partial x_1} a_2 + \dfrac{\partial}{\partial x_2} a_1 \right) u_2(x) = \lambda u_2(x) \end{cases} \qquad (1.2.5)$$

Let

$$\phi_\gamma(x) = \begin{pmatrix} 0 \\ e^{i(\gamma + t, x)} \end{pmatrix}, \qquad \text{where } \gamma \in \Gamma, \ t \in F. \qquad (1.2.6)$$

Then as follows from (1.2.5)

$$\lambda_\gamma = |\gamma + t|^2 - 2(\gamma + t, a) + a^2 - i[\gamma + t, a]. \qquad (1.2.7)$$

Now assume that

$$\phi^\gamma(x) = \begin{pmatrix} e^{i(\gamma + t, x)} \\ 0 \end{pmatrix}, \qquad \text{where } \gamma \in \Gamma, \ t \in F. \qquad (1.2.8)$$

Then from (1.2.5) follows

$$\lambda^\gamma = |\gamma + t|^2 - 2(\gamma + t, a) + a^2 + i[\gamma + t, a]. \qquad (1.2.9)$$

Thus we obtained that eigenvalues and eigenfunctions of the operator $H_0^t(a)$ indeed are λ_γ, λ^γ and $\phi_\gamma(x)$, $\phi^\gamma(x)$ correspondingly.

Let us prove that other eigenvalues of the operator $H_0^t(a)$ in $L_2(F) \times L_2(F)$ are absent. Since $\{\phi_\gamma(x)\}_{\gamma \in \Gamma}$, $\{\phi^\gamma(x)\}_{\gamma \in \Gamma}$ are orthonormal basises in $L_2(F)$, then $\{\phi_\gamma(x)\} \cup \{\phi^\gamma(x)\}$ will be an orthonormal basis in $L_2(F) \times L_2(F)$. As follows from this other eigenvalues of the operator $H_0^t(a)$ are absent. Therefore the spectrum of the $H_0^t(a)$ consists of λ_γ and λ^γ.

And now begin to analyze the operator $H_t(a, V(x))$. To avoid the technical details, let's to begin with, that the potential $V(x)$ is a trigonometric polynomial

$$V(x) = \sum_{\gamma \in Q} V_\gamma e^{i(\gamma, x)}, \tag{1.2.10}$$

where
$$V_\gamma = \int_F V(x) e^{-i(\gamma, x)} dx, \qquad Q = \{\gamma \in \Gamma : V_\gamma \neq 0\}. \tag{1.2.11}$$

Also without loss of generality, we assume that $\int_F V(x) dx = 0$.

From the definition of Q we see immediately that the eigenvalue λ^γ is associated with the eigenvalues $\lambda^{\gamma+\gamma_1}$ and eigenvalue λ_γ is associated with the eigenvalues $\lambda_{\gamma+\gamma_1}$ when (if $\gamma_1 \in Q^{(m)}$) through perturbation $V(x)$ ($V^{(m)}(x)$) where

$$Q^{(m)} = \{\gamma : \gamma = \gamma_1 + \gamma_2 + \ldots + \gamma_p, \gamma_1, \gamma_2, \ldots, \gamma_p \in Q; p \leq m\}$$

Really

$$V(x) \cdot I \cdot \phi^\gamma(x) = \begin{pmatrix} \sum_{\gamma_1} V_{\gamma_1} e^{i(\gamma_1, x)} & 0 \\ 0 & \sum_{\gamma_1} V_{\gamma_1} e^{i(\gamma_1, x)} \end{pmatrix} \begin{pmatrix} e^{i(\gamma+t, x)} \\ 0 \end{pmatrix} = \begin{pmatrix} \sum_{\gamma_1} V_{\gamma_1} e^{i(\gamma+\gamma_1+t, x)} \\ 0 \end{pmatrix}.$$

From this we get A

$$\left| \left(V(x) \cdot I \cdot \phi^\gamma(x), \phi^{\gamma+\gamma_1}(x) \right) \right| = \left| \sum_{\gamma_1 \in Q} V_{\gamma_1} \right| > \left| \lambda^\gamma \right|^{-\alpha}. \tag{1.2.12}$$

Thus λ^γ is associated with $\lambda^{\gamma+\gamma_1}$ by $\gamma_1 \in Q$ (by $\gamma_1 \in Q^{(m)}$) through perturbation $V(x)$ ($V^{(m)}(x)$). By the same way one can show that λ_γ is associated with $\lambda_{\gamma+\gamma_1}$ by $\gamma_1 \in Q$ (by $\gamma_1 \in Q^{(m)}$) through $V(x)$ ($V^{(m)}(x)$).

Now let us estimate the difference $\left| \lambda^\gamma - \lambda^{\gamma+\gamma_1} \right|$ when $\gamma_1 \in Q$.

$$\left| \lambda^\gamma - \lambda^{\gamma+\gamma_1} \right| = \left| |\gamma+t|^2 - 2(\gamma+t, a) + a^2 + i[\gamma+t, a] - |\gamma+\gamma_1+t|^2 + 2(\gamma+\gamma_1+t, a) - a^2 - i[\gamma+\gamma_1+t, a] \right| =$$

$$= \left| |\gamma+t|^2 - |\gamma+\gamma_1+t|^2 + 2(\gamma_1, a) - i[\gamma_1, a] \right| > \left| |\gamma+t|^2 - |\gamma+\gamma_1+t|^2 \right| - \left| i[\gamma_1, a] - 2(\gamma_1, a) \right|.$$

Let γ satisfies the inequality

$$\left| |\gamma+t|^2 - |\gamma+\gamma_1+t|^2 \right| > |\gamma+t|^\alpha \tag{1.2.13}$$

By the elementary calculations we get

$$\left| i[\gamma_1, a] - 2(\gamma_1, a) \right| < C|\gamma_1||a| < M|\gamma+t| < \frac{1}{2}|\gamma+t|^\alpha.$$

These inequalities give

$$\left| \lambda^\gamma - \lambda^{\gamma+\gamma_1} \right| > \frac{1}{2}|\gamma+t|^\alpha \tag{1.2.14}$$

Denote

$$V_\gamma(\alpha) = \left\{ x \in R^2 : \left| \|x\|^2 - |x+\gamma|^2 \right| < |x|^\alpha \right\},$$ (1.2.15)

where $\gamma \in Q^{(m)}$, $\quad Q^{(m)} = \left\{ \gamma : \gamma = \gamma_1 + \gamma_2 + \ldots + \gamma_p, \gamma_1, \gamma_2, \ldots, \gamma_p \in Q; p \le m \right\},$ (1.2.16)

$$V^{(m)}(\alpha) = \bigcup_{\gamma_1 \in Q^{(m)}} V_{\gamma_1}(\alpha).$$ (1.2.17)

Note that the number of the vectors γ included $Q^{(m)}$ is not more than $|Q|^m$

$$\left| Q^{(m)} \right| \le |Q|^m,$$ (1.2.18)

where, $|Q|$ is a number of the vectors γ included Q.

One can easily prove that for any γ, β the eigenvalues λ^γ can never be associated with the eigenvalues λ_β through perturbation $V(x)$ ($V^{(m)}(x)$).

Indeed

$$\left| \left(V(x) \cdot I \cdot \phi^\gamma(x), \phi_\beta(x) \right) \right| = \left| \left(\left(\sum_{\gamma_1} V_{\gamma_1} e^{i(\gamma + \gamma_1 + t, x)} \right), \begin{pmatrix} 0 \\ e^{i(\beta + t, x)} \end{pmatrix} \right) \right| = 0.$$

So eigenvalue λ^γ is m-order multiply non-resonance if and only if the vector $\gamma + t$ does not belong to the set.

From (1.2.18) it follows that if $\rho \to \infty$, then

$$\mu\left(V_{\gamma_1}(\alpha) \cap S_\rho \right) = \mu(S_\rho) \cdot O(\rho^{\alpha-1}), \quad \mu\left(V^{(m)}(\alpha) \cap S_\rho \right) = \mu(S_\rho) \cdot \left[1 + O(\rho^{\alpha-1}) \right].$$ (1.2.19)

Here

$$S_\rho = \left\{ x \in R^2 : |x| = \rho \right\}, \qquad \overline{V}^{(m)}(\alpha) = R^2 \setminus V^{(m)}(\alpha).$$

Thus, $\overline{V}^{(m)}(\alpha)$ is the main part of R^2. This means that m-order non-resonance eigenvalues constitute the majority of all the eigenvalues. And the rest of the eigenvalues will be resonance eigenvalues.

Let us numerate all eigenvalues of the operator $H_t(a, V(x))$ by the pints of the lattice Γ I.e. denote by $\Lambda(\lambda_\gamma)$ or $\Lambda(\lambda^\gamma)$, where $\gamma \in \Gamma$. The set $\left\{ \Lambda(\lambda_\gamma), \Lambda(\lambda^\gamma) : \gamma \in \Gamma \right\}$ constitutes all spectrum of the operator $H_t(a, V(x))$.

By the fulfillment of the condition

$$\sum_{\gamma \in Q} |V_\gamma| < \infty,$$

(it holds for sufficiently smooth periodic function, and for trigonometric polynomial even more) from Theorems 1.1 and 1.4 of the work[6] it follows the theorem.

Theorem 1.2.2. a) If the eigenvalue λ_γ is m-order non-resonance (i.e. $\gamma + t \notin V^{(m)}(\alpha)$), then the corresponding eigenvalue $\Lambda(\lambda_\gamma)$ of the operator $H_t(a, V(x))$ satisfies the asymptotics (1.2.20)

$$\Lambda(\lambda_\gamma) = \lambda_\gamma + \tilde{F}_{m-1}(\lambda_\gamma) + O\left(|\gamma + t|^{-\alpha m}\right), \tag{1.2.20}$$

where $F_1(\lambda_\gamma) = \tilde{F}_1(\lambda_\gamma) = S_1(\lambda_\gamma) = \sum_{\gamma_1} \dfrac{|V_{\gamma - \gamma_1}|}{\lambda_\gamma - \lambda_{\gamma_1}}$, $\tilde{F}_n(\lambda_\gamma) = \sum_{k=1}^{n} S_k\left(\tilde{F}_{m-1}(\lambda_\gamma), \lambda_\gamma\right)$, $n = \overline{2, m-1}$

$$S_n\left(\tilde{F}_{m-1}(\lambda_\gamma), \lambda_\gamma\right) = \sum_{\gamma_1, \gamma_2, \ldots, \gamma_k} \frac{V_{\gamma - \gamma_1} V_{\gamma_2 - \gamma_1} \cdots V_{\gamma_{n-1} - \gamma_n} V_{\gamma_n - \gamma}}{\prod_{s=1,k}\left(\lambda_\gamma + \tilde{F}_{m-1}(\lambda_\gamma) - \lambda_{\gamma_s}\right)} \quad \text{and} \quad \tilde{F}_n(\lambda_\gamma) = O\left(|\gamma + t|^{-\alpha n}\right).$$

б) For any m-order non-resonance eigenvalue λ_γ there exists an eigenvalue Λ_N of the operator $H_t(a, V(x))$ satisfying the asymptotics (1.2.20).

Theorem 1.2.3. a) Let the eigenvalue $\Lambda_N(t)$ of the operator $H_t(a, V(x))$ be such that

$$|\Lambda_N - \lambda_\gamma| > 2M, \qquad |(\Psi_{N,t}, \varphi_\gamma)| > |\gamma + t|^{-\alpha m_0},$$

where $M \geq \sum_\gamma |V_\gamma| \geq \sup_x V(x)$. If λ_γ is m-order non-resonance, then Λ_N satisfies

$$\Lambda_N = \lambda_\gamma + \tilde{F}_{m-m_0-1}(\lambda_\gamma) + O\left(|\gamma + t|^{-\alpha(m-m_0)}\right). \tag{1.2.21}$$

б) Eigenvalue $\Lambda_{k\lambda}$, for which is true

$$\left|(\Psi_{k,t}(h), \varphi_\gamma(h))\right| = \max_N \left\{\left|(\Psi_{k,t}(h), \varphi_\gamma(h))\right|, \left|(\Psi_{k,t}(h), \varphi^\gamma(h))\right|\right\}$$

satisfies to the asymptotics (1.2.41) с $m_0 = \left[\dfrac{n}{2\alpha}\right]$, where $\left[\dfrac{n}{2\alpha}\right]$ -is an integer part of $\dfrac{n}{2\alpha}$, $k \equiv k\lambda$.

Theorem 1.2.4. a) If the eigenvalue λ_γ is singly resonance i.e. $\gamma + t \in V_{\gamma_1}(\alpha)$, then the corresponding eigenvalue of the operator $H_t(a, V(x))$ satisfies to the formula

$$\Lambda_N = \tilde{\lambda}_\gamma + O\left(|\tilde{\lambda}_\gamma|^{\alpha/2}\right), \tag{1.2.22}$$

where $\tilde{\lambda}_\gamma$ is an eigenvalue of the matrix $C = (c_{ij})_{i,j=-p_0, p_0}$, with $c_{ii} = |\lambda_{\gamma + i\gamma_1}|$, $c_{ij} = V_{(i-j)\gamma_1}$.

б) For any singly resonance eigenvalue λ_γ (where $\gamma \in V_{\gamma_1}(\alpha)$) there exists an eigenvalue Λ_N of the operator $H_t(a, V(x))$ from $[\lambda_\gamma - 2M, \lambda_\gamma + 2M]$ satisfying to (1.2.22). Here $M \geq \sum_\gamma |V_\gamma| \geq \sup_x V(x)$.

Now let us prove that $V(x) \in L_2(F)$, i.e. $\|V(x)\|_{L_2(F)} = M$, where

$$V_\gamma = \int_F V(x) e^{-i(\gamma,x)} dx, \quad Q = \{\gamma \in \Gamma : V_\gamma \neq 0\}.$$

From the last immediately follows that for any $M_0 > 0$ there exists n_0 such that

$$\left(\sum_{|\gamma| > n_0} |V_\gamma|^2 \right)^{1/2} < \frac{1}{M_0} . \tag{1.2.23}$$

Consider the polynomial

$$V^*(x) = \sum_{\gamma : |\gamma| < n_0} V_\gamma e^{i(\gamma,x)} .$$

And denote by $V_\rho^{n_0}$ the non-resonance set for the polynomial $V^*(x)$, where

$$V_\rho^{n_0} = R^2 \setminus \bigcup_{|\gamma| < n_0} N_\gamma(\rho).$$

Here $\quad N_\gamma(\rho) = \left\{ x \in R^2 : \left| |x|^2 - |x+\gamma|^2 \right| < \rho \right\}, \ \rho < M_0^2 n_0^4.$

As was defined if $x \in V_\rho^{n_0}$, then for any $|\gamma| < n_0$

$$\left\| |x|^2 - |x+\gamma|^2 - 2(\gamma,a) + [\gamma,a] \right\| > \rho , \tag{1.2.24}$$

i.e.

$$\left| \lambda_x - \lambda_{x+\gamma} \right| > \rho . \tag{1.2.24'}$$

Using (1.2.23) and (1.2.24) we prove.

Theorem 1.2.5. If $\gamma + t \in V_\rho^{n_0}$, then in the $2/M_0$ neighborhood of the number λ_γ there exists at least one eigenvalue $\Lambda_N(t)$ of the operator $H_t(a, V(x))$.

Proof. Suppose that in contrary for all Λ_N $(N = 1, 2, ...)$ is valid

$$\left| \Lambda_N - \lambda_\gamma \right| > 2/M_0 . \tag{1.2.25}$$

The formula

$$(\Lambda_N - \lambda_\gamma)(\Psi_N, \phi_\gamma) = \sum_{\gamma_1} c(\gamma, \gamma_1) \cdot (\Psi_N, \phi_{\gamma_1}) + \sum_{\gamma^1} c(\gamma, \gamma^1) \cdot (\Psi_N, \phi^{\gamma^1}),$$

where

$$c(\gamma, \gamma_1) = (V\phi_\gamma, \phi_{\gamma_1}) \equiv V_{\gamma_1 - \gamma}, c(\gamma, \gamma^1) = (V\phi_\gamma, \phi^{\gamma^1}) \equiv 0, \tag{1.2.26}$$

Rewrite in the form

$$(\Lambda_N - \lambda_\gamma)(\Psi_{k,t}(x), \phi_\gamma(x)) = \sum_{\gamma_1 : |\gamma_1| < n_0} V_{\gamma_1} b_{k, \gamma - \gamma_1} + (\Psi_{k,t}(x), (V(x) - V^*(x)) \phi_\gamma(x)),$$

where $b_{k,\gamma} = \left(\Psi_{k,t}(x), \phi_\gamma(x) \right)$.

Raising both sides squared and summing, we have

$$\sum_{|\Lambda_N - \lambda_\gamma| \le 2M} \left| (\Lambda_N - \lambda_\gamma) b_{k,\gamma} \right|^2 = \sum_k \left| \left(\sum_{\gamma_1; |\gamma_1| < n_0} V_{\gamma_1} b_{k,\gamma-\gamma_1} + \left(\Psi_{k,t}(x), \left(V(x) - V^*(x) \right) \phi_\gamma(x) \right) \right) \right|^2 \le$$

$$\le \sum_k \left(2 \left| \sum_{\gamma_1; |\gamma_1| < n_0} V_{\gamma_1} b_{k,\gamma-\gamma_1} \right|^2 + 2 \left| \left(\Psi_{k,t}(x), \left(V(x) - V^*(x) \right) \phi_\gamma(x) \right) \right|^2 \right). \qquad (1.2.27)$$

Considering (1.2.23) we obtain that

$$\sum_k \left| \left(\Psi_{k,t}(x), \left(V(x) - V^*(x) \right) \phi_\gamma(x) \right) \right|^2 \le \left| V(x) - V^*(x) \right| < \frac{1}{M_0^2},$$

$$\sum_{\gamma_1; |\gamma_1| < n_0} \left| V_{\gamma_1} b_{k,\gamma-\gamma_1} \right|^2 < c \cdot n_1 \cdot \sum_{\gamma_1; |\gamma_1| < n_0} \left| b_{k,\gamma-\gamma_1} \right|^2, \qquad (1.2.28)$$

Here n_1 is a number of the vectors γ_1 involved into the circle S_{n_0} i.e. $n_1 > c \cdot n_0^2$.

Using again (1.2.26) one can get

$$\sum_k \left(\sum_{\gamma_1; |\gamma_1| < n_0} \left| V_{\gamma_1} b_{k,\gamma-\gamma_1} \right|^2 \right) < c \cdot n_1 \cdot \sum_{\gamma_1; |\gamma_1| < n_0} \left(\sum_k \left| b_{k,\gamma-\gamma_1} \right|^2 \right) \le$$

$$\le c \cdot \sum_{\gamma_1; |\gamma_1| < n_0} \left(\sum_k \frac{\left| \left(\Psi_{k,t}(x), V(x) \phi_{\gamma-\gamma_1}(x) \right) \right|^2}{\left| \Lambda_k - \lambda_{\gamma-\gamma_1} \right|^2} \right) \le \frac{c \cdot n_1^2 \cdot M^2}{\min_{\gamma_1} \left(\Lambda_k - \lambda_{\gamma-\gamma_1} \right)^2} \qquad (1.2.29)$$

Now from (1.2.24) and from the inequality

$$\left| \Lambda_k - \lambda_\gamma \right| \le 2M$$

follows that

$$\min_{\gamma_1} \left| \Lambda_k - \lambda_{\gamma-\gamma_1} \right| > \rho - 2M.$$

Therefore (see (1.2.28)) the right hand side of (1.2.27) is less than

$$\frac{1}{M_0^2} + c \cdot n_1^2 \cdot M^2 \cdot (\rho - 2M)^2$$

since

$$\sum_{N; |\Lambda_N(t) - \lambda_\gamma| < 2M} \left| \left(\Psi_{N,t}(x), \phi_\gamma(x) \right) \right| \ge 1 - \frac{1}{4} = \frac{3}{4} \qquad (1.2.30)$$

This inequality is proved in $[6]$ when $V(x)$ is a trigonometric polynomial.

From the other hand the left hand side of (1.2.27) (see (1.2.25) and (1.2.30)) is greater than

$$\left(\frac{2}{M_0}\right)^2 \sum_{k:|\Lambda_k(t)-\lambda_\gamma|<2M} |b_{k,\gamma}|^2 > \frac{4}{M_0^2} \cdot \frac{3}{4} = \frac{3}{M_0^2},$$

which contradicts the condition $\rho > M_0^2 \cdot n_0^4$.

Thus in the $\frac{2}{M_0}$ neighborhood of the number λ_γ there exists at least one eigenvalue $\Lambda_N(t)$ of the operator $H_t(a, V(x))$.

The theorem is proved.

1.3. On a spectrum of the two dimensional periodic Pauli operator (Bethe-Sommerfeld conjecture)

In this chapter we study the Pauli operator $H_t(a, V(x))$ generated in $L_2(F) \times L_2(F)$ by the expession

$$H(a, V(x)) = \left((-i\nabla - a)^2 + V(x)\right) \cdot I + \sigma \cdot B \qquad (1.1.1)$$

and boundary conditions

$$u(x + \omega_j) = e^{2\pi i t_j} u(x) \qquad (1.1.2)$$

where $x \in R^2$, $B = [grad \times a]$ is magnet filed generated by the vector potential $a = (a_1, a_2)$,

$$\sigma = \begin{pmatrix} 1 & 0 \\ 0 & -1 \end{pmatrix}, \qquad I = \begin{pmatrix} 1 & 0 \\ 0 & 1 \end{pmatrix},$$

F is a fundamental domain of some lattice $\Omega = \{m_1\omega_1 + m_2\omega_2 : m_1, m_2 \in Z\}$ i.e. a parallelogram, $t = t_1\gamma^1 + t_2\gamma^2$ and γ^1, γ^2 are biorthogonal to ω_1, ω_2 vectors, F^* is a fundamental domain of the lattice $\Gamma = \{n_1\gamma^1 + n_2\gamma^2 : n_1, n_2 \in Z\}$, $V(x)$ - periodic smooth enough function. Without loss of generality, we assume that. $\mu(F) = 1$.

Here we shall prove that if $V(x)$ is a trigonometric polynomial or a continuous function, the number of gaps in the spectrum of two-dimensional Pauli operator $H(a, V(x))$ is limited. So Bethe-Sommerfeld conjecture is true.

In Section 2 To the eigenvalue $|\gamma + t|^2 - 2(\gamma + t, a) + a^2 \pm [\gamma + t, a]$, where $\gamma \in \Gamma$, $t \in F$ of the unperturbed Pauli operator, we assign a proper eigenvalue (it is denoted by $\Lambda_{k(\gamma+t)}$) the operator $H_t(a, V(x))$ according to the rule

$$\left|\left(\Psi_{k,t}(x), e^{i(\gamma+t, x)}\right)\right| = \max_N \left|\left(\Psi_{N,t}(x), e^{i(\gamma+t, x)}\right)\right|.$$

It is proved that $\Lambda_{k(\gamma+t)}$ satisfies to the formula

$$\Lambda_{k(\gamma+t)} = |\gamma+t|^2 - 2(\gamma+t,a) + a^2 \pm [\gamma+t,a] + \tilde{F}_{m-1}(\gamma+t) + O\left(|\gamma+t|^{[\mathcal{Y}_2]-m\alpha}\right), \quad \gamma+t \in \tilde{V}^m(\alpha);$$

$$\Lambda_{k(\gamma+t)} = \lambda(\gamma+t) + O\left(|\gamma+t|^{-\alpha}\right), \qquad \gamma+t \in V^{(m)}(\alpha) = R^2 \setminus \tilde{V}^{(m)}(\alpha),$$

where $V^{(m)}(\alpha) = \bigcup_\gamma V_\gamma(\alpha)$, $V_\gamma(\alpha) = \left\{x \in R^2 : \left||x|^2 - |x+\gamma|^2\right| < |x|^\alpha\right\}$

In $[6]$ using the asymptotic formulas for the eigenvalues of the Schrödinger operator the set B with the following properties is constructively constructed:

I. By $\gamma+t \in B$ eigenvalue $\Lambda_{k(\gamma+t)}$ of the Schrödinger operator is simple, and the corresponding eigenfunction $\Psi_{k(\gamma+t)}$ is close to $e^{i(\gamma+t,x)}$ - the eigenfunction of the Laplace operator

$$\left|\left(\Psi_{k(\gamma+t)}, e^{i(\gamma+t,x)}\right)\right| = 1 + O\left(|\gamma+t|^{-\alpha}\right), \qquad \textbf{(III)}$$

moreover, satisfies the asymptotics $\Psi_{k(\gamma+t)} = e^{i(\gamma+t,x)} + \Phi_1(x) + ... + \Phi_{m-1}(x) + O\left(|\gamma+t|^{-m\alpha}\right)$

,

where $\Phi_s(x)$ is explicitly determined and has an order $O\left(|\gamma+t|^{-s\alpha}\right)$, при $s = 1, 2, ..., m-1$.

Thus the speed, current and momentum is calculated corresponding to the state $\Psi_{k(\gamma+t)}$ by $\gamma+t \in B$.

II. The set B constitutes the main part of R^2:

$$\mu(B \cap S_\rho) = (1 + O(\rho^{-\alpha}))\mu(S_\rho)$$

And for large ρ involves the intervals

$$A_b(\delta) = \left\{b + \tau\frac{b}{|b|} : \tau \in [-\delta, \delta]\right\}, \qquad (a \in S_\rho = \{x \in R^2 : |x| = \rho\}, \delta \approx \rho)$$

From the asymptotics of the eigenvalues of the Schrodinger operator follows that

a) $\Lambda_{k\left(b-\tau\frac{b}{|b|}\right)} < \rho^2 - 2(\rho,a) + a^2 + [\rho,a] \equiv \lambda^\rho$, $\Lambda_{k\left(b-\tau\frac{b}{|b|}\right)} > \lambda^\rho$,

б) As follows from (III) $\Lambda_{k(\gamma+t)}$ continuously depends on $\gamma+t$ by $\gamma+t \in B$, and so by $\gamma+t \in A_b(\delta) \subset B$.

It follows from the above considerations by large ρ there exit eigenvalues $\Lambda_{k(\gamma+t)}$ (where $\gamma+t \in A_b(\delta) \subset B$) coinciding with the number λ^ρ. Thus, it is possible to prove that the constant-energy surface contains the subsets $\{\gamma+t \in B : \Lambda_{k(\gamma+t)} = \lambda^\rho\}$ of measure is asymptotically $(\rho \to \infty)$

close to the measures of the sphere S_ρ. Since $\{\gamma + t \in B : \Lambda_{k(\gamma+t)} = \lambda^\rho\} \neq \emptyset$ for large ρ, it follows the validity of the Bethe – Sommerfeld conjecture.

Note that the main difficulty ism a construction of the set B with the properties I, II. Indeed condition I is equivalent to the equality

$$\sum_{\tilde{\gamma} \in \Gamma / \gamma} \left| \left(\tilde{\Psi}_{k(\gamma+t)}, e^{i(\gamma+t,x)} \right) \right| = O\left(|\gamma + t|^{-2\alpha} \right), \tag{IV}$$

where $\tilde{\Psi}_{k(\gamma+t)}$ is any eigenfunction corresponding to the eigenvalue $\Lambda_{k(\tilde{\gamma}+t)}$. That is why in $[6]$ in some cases the set $B_\rho(\delta)$ (indeed B, where $B = \bigcup B_\rho(\delta)$) is constructed considering (IV), and in other ones considering that $\Lambda_{k(\tilde{\gamma}+t)}$ should be simple when $\gamma + t \in B$.

Now we transfer the results obtained in $[2]$ to the case of Pauli operator in parallelogram.

Assume that ρ **is** a sufficiently large number. To prove the inclusion λ^ρ in the spectrum of the operator (1.1.1) it is enough to find an interval $A_b(\delta)$ that satisfies the conditions a) and b). Let

$$A_b(\delta) = \left\{ b + \tau \frac{b}{|b|} : \tau \in [-\delta, \delta] \right\}, \ b \in S \ , \delta \approx \rho^{-1}, b \in V^m(\alpha)$$

It is easy to see that the condition a is satisfied. Really considering (II)

$$\Lambda_{k\left(b+\delta\frac{b}{|b|}\right)} = \left| b + \delta\frac{b}{|b|} \right|^2 - 2\left(b + \delta\frac{b}{|b|}, a\right) + a^2 + \left[b + \delta\frac{b}{|b|}, a\right] + O(\rho^{-\alpha}) > \lambda^\rho,$$

$$\Lambda_{k\left(b-\delta\frac{b}{|b|}\right)} = \left| b - \delta\frac{b}{|b|} \right|^2 - 2\left(b - \delta\frac{b}{|b|}, a\right) + a^2 + \left[b - \delta\frac{b}{|b|}, a\right] + O(\rho^{-\alpha}) < \lambda^\rho, \tag{1.3.1}$$

since $2\delta\rho \approx 1$.

To check the validity of the condition б) we use the lemma.

Lemma 1: If for $\gamma + t \in A_b(\delta)$ the eigenvalue $\Lambda_{\kappa(\gamma+t)}$ is simple and

$$\left| \left(\Psi_{k,t}(x), \varphi^\gamma(x) \right) \right|^2 > \frac{1}{2} \tag{1.3.2}$$

then the index κ in the interval $A_b(\delta) = \left\{ b + \tau \frac{b}{|b|} : \tau \in [-\delta, \delta] \right\}$ does not depend on τ, i.e.

$\kappa\left(b + \tau\frac{b}{|b|} \right) \equiv const$ by $\tau \in [-\delta, \delta]$ and consequently the function $\Lambda_{\kappa\left(b+\tau\frac{b}{|b|}\right)}$ is continuous in the

interval $[-\delta, \delta]$.

Proof: Let $\gamma + t_0 \in A_b(\delta)$. From the simplicity condition of the eigenvalue $\Lambda_{\kappa(\gamma+t_0)} \equiv \Lambda_\kappa(t_0)$ follows that in the enough small neighborhood u_{t_0} of the pint t_0 the eigenvalue $\Lambda_\kappa(t)(\kappa = \kappa(\gamma+t))$

is simple, the function $\left|\left(\Psi_{k,t}(x),\varphi^{\gamma}(x)\right)\right|$ is continuous with respect to t and so

$$\left|\left(\Psi_{k,t}(x),\varphi^{\gamma}(x)\right)\right|^2 > \frac{1}{2} \quad \text{by } t \in u_{t_0}, \quad \text{i.e. } k(\gamma+t) = k(\gamma+t_0) \text{ by } t \in u_{t_0}.$$

It is obvious that $k(\gamma+t)$ is constant in each connected compact $V \subset A_b(\delta)$, since the compact V has finite cover $\{u_{t_i}\}_i$, and in each u_{t_i} and their intersection $\kappa(\gamma+t)$ is constant.

Theorem 1.3.1: If for $\gamma+t \in A_b(\delta)$

$$\sum_{\tilde{\gamma}:\tilde{\gamma}\in\Gamma/\gamma} \left|\left(\tilde{\Psi}_{k(\gamma+t)}(x),\varphi^{\gamma}(x)\right)\right|^2 < \frac{1}{2} \tag{1.3.3}$$

Holds true, where $\tilde{\Psi}_{k(\gamma+t)}(x)$ is any normalized eigenfuction corresponding to the eigenvalue $\Lambda_{\kappa(\tilde{\gamma}+t)}$, then the point λ^{ρ} belong to the spectrum of the operator $H(a,V(x))$.

Proof: It is sufficient to show that the conditions of Lemma 1 are satisfied. Let $\Lambda_{\kappa(\gamma+t)}$ be a non-simple eigenvalue of the operator $H(a,V(x))$. Then there exist two orthonormal eigenfuctions $\Psi^1_{\kappa,t}$, $\Psi^2_{\kappa,t}$ such that

$$\sum_{\tilde{\gamma}\neq\gamma}\left|\left(\Psi^s_{\kappa,t}(x),\varphi^{\tilde{\gamma}}(x)\right)\right|^2 < \frac{1}{2}, \quad s = 1,2.$$

From this considering Perceval's equality we obtain

$$\left|\left(\Psi^s_{\kappa,t}(x),\varphi^{\gamma}(x)\right)\right|^2 > \frac{1}{2}, \quad s = 1,2.$$

Using this and normality of $\varphi^{\gamma}(x)$ we come to the contradiction

$$1 = \left\|\varphi^{\gamma}(x)\right\| \geq \sum_{s=1}^{2}\left|\left(\Psi^s_{\kappa,t}(x),\varphi^{\gamma}(x)\right)\right|^2 > 1.$$

Thus $\Lambda_{\kappa(\gamma+t)}$ is simple and (1.3.2) is satisfied.

Thus the proof of the Bethe-Sommerfeld conjecture is reduced to the finding of the interval $A_b(\delta)$ (where $a \in S_\rho, \delta \approx \rho^{-1}$) such that by $\gamma+t \in A_b(\delta)$ holds true (1.3.3).

Let us estimate the sum in the right hand side of (1.3.3). For this purpose we divide the set $\{\tilde{\gamma}+t : \tilde{\gamma} \in \Gamma \setminus \{\gamma\}\}$ by the parts J_1, J_2, J_3, where

$$J_1 = \left\{\tilde{\gamma}+t : \tilde{\gamma}\in\Gamma, \tilde{\gamma}+t \notin K_\rho\left(\frac{M}{\rho}\right)\right\}, \quad \text{where } K_\rho\left(\frac{M}{\rho}\right) = \left\{x : \|x\| - \rho < \frac{M}{\rho}\right\};$$

$$J_2 = \left\{\tilde{\gamma}+t : \tilde{\gamma}\in\Gamma, \tilde{\gamma}+t \in K_\rho\left(\frac{M}{\rho}\right) \cap \left(R^2 \setminus V^m(\alpha)\right)\right\};$$

$$J_1 = \left\{ \tilde{\gamma} + t : \tilde{\gamma} \in \Gamma, \tilde{\gamma} + t \in K_\rho \left(\frac{M}{\rho} \right) \cap V^m(\alpha) \right\};$$

$$M > 2\sum_{\gamma \in \Gamma} |V_\gamma| \geq 2\sup|V(x)|, \qquad M > 3\rho\delta.$$

By this way the sum in the right hand side of (1.3.3) also is divided by three sums $\Sigma_1, \Sigma_2, \Sigma_3$, where Σ_i is a sum over J_i, $i = 1,2,3$.

We estimate each sum separately. By definition of J_1 for $\tilde{\gamma} + t \in J_1$ we have

$$\left| |\tilde{\gamma} + t|^2 - \rho^2 \right| > 2M \Rightarrow \left| \lambda^{\tilde{\gamma}} - \lambda^\rho \right| > 2M$$

Considering (II)

$$\left| \Lambda_{\kappa(\gamma+t)} - \lambda^\gamma \right| = O\left(|\gamma|^{-\alpha} \right), \quad \text{where } \left| \lambda^\gamma - \lambda^\rho \right| < \frac{5}{2}\rho\delta$$

for $\gamma + t \in A_b(\delta) \subset K_\rho(\delta) \cap \tilde{V}^m(\alpha)$, i.e. $\left| \Lambda_{\kappa(\gamma+t)} - \lambda^\rho \right| < M.$

Therefore

$$\left| \Lambda_{\kappa(\gamma+t)} - \lambda^{\tilde{\gamma}} \right| > M \quad \text{for } \tilde{\gamma} + t \in J_1, \; \gamma + t \in K_\rho(\delta), \; M > 3\rho\delta.$$

Using the last formulas and

$$\left(\Lambda_{\kappa(\gamma+t)} - \lambda^\gamma \right)\left(\Psi_{\kappa(\gamma+t)}(x), \varphi^\gamma(x) \right) = \left(\Psi_{\kappa(\gamma+t)}(x), V(x)\varphi^\gamma(x) \right),$$

we get

$$\Sigma_1 \equiv \sum_{\tilde{\gamma}:\tilde{\gamma}+t \in J_1} \left| \left(\tilde{\Psi}_{\kappa(\gamma+t)}, \varphi^\gamma \right) \right|^2 < \frac{(\sup V(x))^2}{M^2} < \frac{1}{4}, \text{ for } \gamma + t \in K_\rho(\delta). \qquad (1.3.4)$$

Further considerations are similar to[6]. First the subset $B_\rho(\delta) \subset K_\rho(\delta)$ is constructed such that for $\gamma + t \in B_\rho(\delta)$ the sum Σ_2 be small enough. Then the subset $\tilde{B}_\rho(\delta) \subset B_\rho(\delta)$ is constructed for which by $\gamma + t \in \tilde{B}_\rho(\delta)$ enough small the sum Σ_3. Then it is shown that the interval $A_b(\delta)$ is involved in $\tilde{B}_\rho(\delta)$. Consequently by $\gamma + t \in A_b(\delta) \subset \tilde{B}_\rho(\delta) \subset B_\rho(\delta) \subset \left(K_\rho(\delta) \cap \tilde{V}^m(\alpha) \right)$ all the sums $\Sigma_1, \Sigma_2, \Sigma_3$ will be small i.e. (1.3.3) will be satisfied.

The set $B_\rho(\delta)$ is constructed as follows:

Let $\qquad \delta \approx \rho^{-1}, \quad \varepsilon = 17\rho\delta, \quad \alpha_1 = const, \alpha_1 < \frac{1}{2}. \qquad (1.3.5)$

take

$$B_\rho(\delta) = \bigcup_{b \in S'_\rho} u_b(\delta), \qquad (1.3.6)$$

where S'_ρ is a part of sphere S_ρ that does not contain the δ-neighborhood of the m-order non-resonance set $V^m(\alpha)$ and the set

$$\Omega_\rho(\varepsilon) = \bigcup_{\gamma \in \Gamma}(N_\gamma(\varepsilon) \cap S_\rho), \qquad (1.3.7)$$

Here: $N_\gamma(\varepsilon) = \left\{ x : \left| |x|^2 - |x-\gamma|^2 \right| < \varepsilon \right\}$.

In [5] is proved that $\mu(\Omega_\rho(\varepsilon)) = C_\Gamma \varepsilon \mu(S_\rho)$.

If $\varepsilon < \dfrac{2\alpha_1}{C_\Gamma}$ then $\mu(\Omega_\rho(\varepsilon)) < 2\alpha_1 \mu(S_\rho)$ and so

$$\mu(S'_\rho) > (1-\alpha_1)\mu(S_\rho) > \frac{1}{2}\mu(S_\rho). \qquad (1.3.8)$$

Lemma2: There exists $\rho_0 \equiv \rho_0(V(x))$ such that by $\rho > \rho_0$ the set $B_\rho(\delta)$ has the following properties:

1) If $x \in B_\rho(\delta)$ then for any $\gamma \in \Gamma \setminus \{0\}$

$$\left| |x|^2 - |x-\gamma|^2 \right| > \frac{\varepsilon}{2}; \qquad (1.3.9)$$

2) If $x \in B_\rho(\delta)$ then for any $\gamma \in \Gamma \setminus \{0\}$, $x - \gamma \notin B_\rho(\delta)$;

3) $B_\rho(\delta) \subset \tilde{V}^m(\alpha)$, $\mu(B_\rho(\delta)) \approx 1$.

Proof: Let us rewrite the expression $\left| |x|^2 - |x-\gamma|^2 \right|$ in the form

$$|x|^2 - |x-\gamma|^2 = \left(|x|^2 - |b|^2 \right) + \left(|b|^2 - |b-\gamma|^2 \right) + \left(|b-\gamma|^2 - |x-\gamma|^2 \right), \qquad (1.3.10)$$

where b is the center of the ball which contains the point x from $B_\rho(\delta)$. From the relation $|b| = \rho$, $\delta \approx \rho^{-1}$, $x \in u_b(\delta)$ follows that $\left| |x| - |b| \right| < \delta$, $|x| + |b| < 2\delta + 1$, $\left| |x-\gamma| - |b-\gamma| \right| \le |x-b| < \delta$. It is also obvious that $|\gamma| < 2\rho + 1$ (в (1.3.9)) otherwise (1.3.9) will be satisfied automatically.

Using these inequalities we obtain

$$\left| |x|^2 - |b|^2 \right| = \left| |x| - |b| \right|(|x| + |b|) < (2\rho + 1)\delta,$$

$$\left| |x-\gamma|^2 - |b-\gamma|^2 \right| = \left| |x-\gamma| - |b-\gamma| \right|(|x-\gamma| + |b-\gamma|) < (6\rho + 3)\delta. \qquad (1.3.11)$$

From the other hand, from the definition of $B_\rho(\delta)$, $b \notin N_\gamma(\varepsilon)$ i.e.

$$\left| |b|^2 - |b-\gamma|^2 \right| > \varepsilon. \qquad (1.3.12)$$

Therefore from (1.3.10), (1.3.11), (1.3.12) follows that (1.3.9) is true considering $\varepsilon = 17\rho\delta$ (see (1.3.5)).

If to assume that, x and also $x - \gamma \in B_\rho(\delta)$ then то it means that both these pints are in the δ -neighborhood of the sphere S_ρ, and so

$$\left\|x\right\| - \left|x - \gamma\right\| < 2\delta, \qquad \left\|x\right\| + \left|x - \gamma\right\| < 2\rho + 1.$$

From this follows that

$$\left\| |x|^2 - |x - \gamma|^2 \right\| = \left\||x| - |x - \gamma|\right\|\left(|x| + |x - \gamma|\right) < (2\rho + 1)2\delta < \frac{\varepsilon}{2}.$$

That contradicts to the first property of $B_\rho(\delta)$. Consequently, the second property is also true.

The third property follows from (1.3.8) and $\delta \approx \rho^{-1}$ if to consider the definition of $B_\rho(\delta)$

since $\qquad \mu\left(B_\rho(\delta)\right) \approx \mu\left(S'_\rho\right)\delta \approx \rho\rho^{-1} = 1.$

Now we show that by $\gamma + t \in B_\rho(\delta)$ is valid $\Sigma_2 << 1$.

The number of the vectors $\tilde{\gamma} + t$ (where $\tilde{\gamma} \in \Gamma$) belonging to J_2 is less than ρ^2. Therefore if we show that

$$\left|\left(\widetilde{\Psi}_{\kappa(\gamma+t)}, \varphi^\gamma(x)\right)\right| < \rho^{-2}, \text{ for } \gamma + t \in J_2, \tag{2.3.13}$$

then we get

$$\Sigma_2 \equiv \sum_{\tilde{\gamma}:\tilde{\gamma}+t\in J_2} \left|\left(\widetilde{\Psi}_{\kappa(\gamma+t)}, \varphi^\gamma\right)\right|^2 < \rho^{-4}\rho^2 << 1. \tag{2.3.14}$$

The inequality (2.3.13) is proved by prove by contradiction. Let

$$\left|\left(\widetilde{\Psi}_{\kappa(\gamma+t)}, \varphi^\gamma(x)\right)\right| > \rho^{-2}, \qquad \text{где } \tilde{\gamma} + t \in B_\rho(\delta) \subset V^m(\alpha).$$

Then by the asymptotics

$$\left|\Lambda_{\kappa(\gamma+t)} - |\tilde{\gamma} + t|^2\right| = O\left(\rho^{-\alpha}\right) \Rightarrow \left|\Lambda_{\kappa(\gamma+t)} - \lambda^{\tilde{\gamma}}\right| = O\left(\rho^{-\alpha}\right).$$

From the last and (II) one can obtain

$$\left|\lambda^\gamma - \lambda^{\tilde{\gamma}}\right| << \varepsilon \Rightarrow \left\||\gamma + t|^2 - |\tilde{\gamma} + t|^2\right\| << \varepsilon, \quad \text{где } \gamma + t \in B_\rho(\delta), \tilde{\gamma} + t - (\gamma + t) \in \Gamma,$$

That contradicts to the property of the set $B_\rho(\delta)$.

Consequently (1.3.14) is valid for $\gamma + t \in B_\rho(\delta)$.

Now we construct the subset $\widetilde{B}_\rho(\delta) \subset B_\rho(\delta)$ such that b7y $\gamma + t \in \widetilde{B}_\rho(\delta)$ take place the inequality $\Sigma_2 << 1$, i.e. (1.3.3) is valid. In the considered two dimensional case a set A' may be omit from $B_\rho(\delta)$ such that for $\gamma + t \in B_\rho(\delta) \backslash A'$ the set J_3 be empty i.e. Σ_3 is absent. Taking

$$A' = \{x + \gamma : x \in A, \gamma \in \Gamma\}, \qquad A = K_\rho\left(\frac{M}{\rho}\right) \cap V^m(\alpha), \qquad \widetilde{B}_\rho(\delta) = B_\rho(\delta) \backslash A'.$$

we see that if $\gamma + t \in \tilde{B}_\rho(\delta)$ (i.e. $\gamma + t \notin A'$), then for any $\gamma_1 \in \Gamma$

$$\gamma_1 + t \notin A \qquad (\text{т.е. } J_3 = \emptyset). \tag{1.3.15}$$

Otherwise will be valid

$$\gamma + t = \gamma_1 + t + (\gamma - \gamma_1) \in A'.$$

Thus if

$$\gamma + t \in \tilde{B}_\rho(\delta) \subset B_\rho(\delta) \subset K_\rho(\delta), \text{ then } \Sigma_1 < \frac{1}{4}, \Sigma_2 \ll 1, \Sigma_3 = 0.. \tag{1.3.16}$$

From this follows that $\Sigma = \Sigma_1 + \Sigma_2 + \Sigma_3 < \frac{1}{2}$ i.e. (1.3.3) is satisfied.

Now we need only to show that $\tilde{B}_\rho(\delta)$ contains the interval $A_b(\delta)$.

Theorem 1.3.2: The set $\tilde{B}_\rho(\delta)$ contains the intervals $A_b(\delta)$, where $b \in S_\rho$, $\delta \approx \rho^{-1}$.

Proof: This fact directly follows from three lemmas below proved in $[26]$.

Lemma: The inequality $\mu(u_A(2\delta)) < \mu(B_\rho(\delta))$ is valid where

$$u_A(2\delta) \equiv \{x \in R^2 : \rho(x, A) \le 2\delta\}. \tag{1.3.17}$$

Лемма: Let $u'_A(2\delta) = \{\gamma + x : x \in u_A(2\delta), \gamma \in \Gamma\}$. Then $\mu(u'_A(2\delta)) \le \mu(u_A(2\delta))$.

Лемма: If the inequality

$$\mu(u'_A(2\delta)) < \mu(B_\rho(\delta)), \tag{1.3.18}$$

Holds true then there exists the interval $A_b(\delta)$ belonging to $B_\rho(\delta)$.

Theorem 1.3.3 (Bethe-Sommerfeld conjecture): If $V(x)$ is a trigonometric polynomial then the number of the gaps in the spectrum of the operator $H(a, V(x))$ is limited.

Proof: By the theorem 1.3.2 the interval $A_b(\delta)$ ($b \in S_\rho$) exists such that $A_b(\delta) \subset \tilde{B}_\rho(\delta)$. Therefore, by (1.3.16) the inequality (1.3.3) is true for $\gamma + t \in A_b(\delta)$. From this by Theorem 1 we get that $\lambda^\rho \in \sigma(H(a, V(x)))$, if ρ is large enough.

Let the potential $V(x)$ be a continuous function. By the Weierstrass theorem for any constant c ($c < \rho\delta < 1$) the polynomial $\tilde{V}(x)$ exists such that $|V(x) - \tilde{V}(x)| < c$.

Denote by $\Lambda_\kappa^{V(x)}(t)$ ($\Lambda_\kappa^{\tilde{V}(x)}(t)$) eigenvalue of the Pauli operator with potential $V(x)$ ($\tilde{V}(x)$). By the general perturbation theory

$$\left|\Lambda_\kappa^{V(x)}(t) - \Lambda_\kappa^{\tilde{V}(x)}(t)\right| < c. \tag{1.3.19}$$

Considering (1.3.1) we have

$$\Lambda_{\kappa\left(b+\delta\frac{b}{|b|}\right)} > \lambda^\rho + \delta\rho \quad \text{and} \quad \Lambda_{\kappa\left(b-\delta\frac{b}{|b|}\right)} < \lambda^\rho - \delta\rho.$$

Moreover by Lemmaм 1 the indexes $\kappa\left(b+\delta\frac{b}{|b|}\right)$ and $\kappa\left(b-\delta\frac{b}{|b|}\right)$ are the same. Other words there

exists such index κ that

$$\left\{\Lambda_\kappa^{\tilde{V}(x)}(t):t\in F\right\}\supset\left[\lambda^\rho-\rho\delta,\lambda^\rho+\rho\delta\right].$$

From this and (19) follows that $\lambda^\rho\in\left\{\Lambda_\kappa^{\tilde{V}(x)}(t):t\in F\right\}$, since $c<\delta\rho$. Thus the following theorem is proved.

Theorem1.3.4 (Bethe-Sommerfeld conjecture): If $V(x)$ is a continuous funftion, then the number of the gaps in the spectrum of the operator $H(a,V(x))$ is finite.

Moreover it is obvious that the Theorem 1.3.4 is valid for the potentials by which the eigenvalue $\Lambda_\kappa^{V(x)}(t)$ satisfies to the condition:

There exists a continuous function $V_1(x)$ such that

$$\left|\Lambda_\kappa^{V(x)}(t)-\Lambda_\kappa^{V_1(x)}(t)\right|<c, \qquad \text{where} \quad c<\delta\rho\approx1.$$

It is unknown, however, whether this inequality is true for any function $L_2(F)$.

Thus we constructed a set $B=\bigcup_\rho B_\rho(\delta)$ with properties I, II. It is obvious that in the

inequality (1.3.8) instead of $\frac{1}{2}$ may be taken $O(\rho^{-\alpha})$ ($\frac{1}{2}$ was taken only for simplicity), and

$$\mu(B\cap S_\rho)=(1+O(\rho^{-\alpha}))\mu(S_\rho) \tag{1.3.20}$$

Thus the set B constitutes the main part (in the sense of (1.3.20)) of the plane R^2, and therefore the set of vectors $b\in S_\rho$, $\gamma+t\in A_b(\delta)$ the conditions a) and б) are satisfied, constitutes the main part of the sphere S_ρ.

Theorem 1.3.5: The set B, for which the eigenvalue $\Lambda_{\kappa(\gamma+t)}$ is simple and (III) is valid, constitutes the main part of R^2 (in the sense of (1.3.20)). Isoenergetic surface corresponding to the energy λ^ρ contains the set

$$A(\rho)=\left\{\gamma+t\in B:\Lambda_{\kappa(\gamma+t)}=\lambda^\rho\right\},$$

The measure of which ρ is close to the measure of S_ρ:

$$\mu(A(\rho))=(1+O(\rho^{-\alpha}))\mu(S_\rho).$$

PART 2

Asymptotic formulas for the eigenvalues Pauli operator
in the parallelepiped

In this chapter we consider operator $H_t(a,V(h))$ generated in $L_2(F) \times L_2(F)$ by the expression

$$H(a,V(h)) = ((-i\nabla - a)^2 + V(h)) \cdot I + \sigma \cdot B \qquad (2.1.1)$$

and boundary conditions

$$u(h + \omega_j) = e^{2\pi i t_j} u(h), \quad j = 1,3,$$

where $h = (x,y,z) \in R^3$, $B = [\nabla,a]$ is a magnet field generated by the vector potential $a = (a_1,a_2,a_3), B = (B_1,B_2,B_3)$

$$I = \begin{pmatrix} 1 & 0 \\ 0 & 1 \end{pmatrix}, \quad B_1 = \begin{vmatrix} \dfrac{\partial}{\partial y} & \dfrac{\partial}{\partial z} \\ a_2 & a_3 \end{vmatrix}, \quad B_2 = \begin{vmatrix} \dfrac{\partial}{\partial z} & \dfrac{\partial}{\partial x} \\ a_3 & a_1 \end{vmatrix}, \quad B_3 = \begin{vmatrix} \dfrac{\partial}{\partial x} & \dfrac{\partial}{\partial y} \\ a_1 & a_2 \end{vmatrix},$$

$$\sigma_1 = \begin{pmatrix} 0 & 1 \\ 1 & 0 \end{pmatrix}, \quad \sigma_2 = \begin{pmatrix} 0 & -i \\ i & 0 \end{pmatrix}, \quad \sigma_3 = \begin{pmatrix} 1 & 0 \\ 0 & -1 \end{pmatrix},$$

F is a fundamental domain for some lattice $\Omega = \{m_1\omega_1 + m_2\omega_2 + m_3\omega_3 : m_1,m_2,m_3 \in Z\}$, i.e. parallelogram, $t = t_1\gamma^1 + t_2\gamma^2 + t_3\gamma^3$, $\gamma^1,\gamma^2,\gamma^3$ are biorthogonal to $\omega_1,\omega_2,\omega_3$ vectors, F^*- fundamental domain of the lattice $\Gamma = \{n_1\gamma^1 + n_2\gamma^2 + n_3\gamma^3 : n_1,n_2,n_3 \in Z\}$, $V(h)$ is periodic, smooth enough function.

Considering (2.1.3) and (2.1.4) the operator $H(a,V(h))$ may be written in the form

$$H(a,V(h)) = \begin{pmatrix} -\Delta + 2i(\nabla,a) + a^2 + V(h) + [\nabla,a]_z & [\nabla,a]_x - i[\nabla,a]_y \\ [\nabla,a]_x + i[\nabla,a]_y & -\Delta + 2i(\nabla,a) + a^2 + V(h) - [\nabla,a]_z \end{pmatrix}.$$

In this part this operator is investigated by two schemes. In the first one we divide $H(a,V(h))$ by the following parts

$$H(a,V(h)) = \begin{pmatrix} -\Delta + 2i(\nabla,a) + a^2 + [\nabla,a]_z & [\nabla,a]_x - i[\nabla,a]_y \\ [\nabla,a]_x + i[\nabla,a]_y & -\Delta + 2i(\nabla,a) + a^2 - [\nabla,a]_z \end{pmatrix} + \begin{pmatrix} V(h) & 0 \\ 0 & V(h) \end{pmatrix}.$$

In the second scheme we do it as follows

$$H(a,V(h)) = \begin{pmatrix} -\Delta + 2i(\nabla,a) + a^2 + [\nabla,a]_z & 0 \\ 0 & -\Delta + 2i(\nabla,a) + a^2 - [\nabla,a]_z \end{pmatrix} + \begin{pmatrix} V(h) & [\nabla,a]_x - i[\nabla,a]_y \\ [\nabla,a]_x + i[\nabla,a]_y & V(h) \end{pmatrix}$$

In the first and second schemes first part will be considered as the unperturbed operator and the second part as a perturbation. The first scheme is a simpler form. However, it is applicable only in the case when $a = const$, the second scheme is more general and can be applied when $a = a(h)$. In the first section we obtain asymptotic formulas of order $O(1)$ for some series of the eigenvalues.

In the second section developing these results, we obtain asymptotic formulas of high order for some series of eigenvalues.

In the third section, all the results are summarized for the Pauli operator.

2.1. Asymptotic formulas of order $O(1)$ for the eigenvalues
periodic Pauli operator in the parallelepiped

In this chapter we consider operator $H_t(a, V(h))$ generated in $L_2(F) \times L_2(F)$ by the expression

$$H(a, V(h)) = \left((-i\nabla - a)^2 + V(h)\right) \cdot I + \sigma \cdot B \qquad (2.1.1)$$

and boundary conditions

$$u(h + \omega_j) = e^{2\pi i t_j} u(h) \qquad j = 1,3 , \qquad (2.1.2)$$

where $h = (x, y, z) \in R^3$, $B = [\nabla, a]$ is a magnet field generated by the vector potential $a = (a_1, a_2, a_3)$, $B = (B_1, B_2, B_3)$

$$I = \begin{pmatrix} 1 & 0 \\ 0 & 1 \end{pmatrix}, \quad B_1 = \begin{vmatrix} \dfrac{\partial}{\partial y} & \dfrac{\partial}{\partial z} \\ a_2 & a_3 \end{vmatrix}, \quad B_2 = \begin{vmatrix} \dfrac{\partial}{\partial z} & \dfrac{\partial}{\partial x} \\ a_3 & a_1 \end{vmatrix}, \quad B_3 = \begin{vmatrix} \dfrac{\partial}{\partial x} & \dfrac{\partial}{\partial y} \\ a_1 & a_2 \end{vmatrix}, \qquad (2.1.3)$$

$$\sigma_1 = \begin{pmatrix} 0 & 1 \\ 1 & 0 \end{pmatrix}, \quad \sigma_2 = \begin{pmatrix} 0 & -i \\ i & 0 \end{pmatrix}, \quad \sigma_3 = \begin{pmatrix} 1 & 0 \\ 0 & -1 \end{pmatrix}, \qquad (2.1.4)$$

F is a fundamental domain for some lattice $\Omega = \{m_1\omega_1 + m_2\omega_2 + m_3\omega_3 : m_1, m_2, m_3 \in Z\}$, i.e. parallelepiped, $t = t_1\gamma^1 + t_2\gamma^2 + t_3\gamma^3$, $\gamma^1, \gamma^2, \gamma^3$ are biorthogonal to $\omega_1, \omega_2, \omega_3$ vectors, F^*-fundamental domain of the lattice $\Gamma = \{n_1\gamma^1 + n_2\gamma^2 + n_3\gamma^3 : n_1, n_2, n_3 \in Z\}$, $V(h)$ is periodic, smooth enough function.

Considering (2.1.3) and (2.1.4) the operator $H(a, V(h))$ may be written in the form

$$H(a, V(h)) = \begin{pmatrix} -\Delta + 2i(\nabla, a) + a^2 + V(h) + [\nabla, a]_z & [\nabla, a]_x - i[\nabla, a]_y \\ [\nabla, a]_x + i[\nabla, a]_y & -\Delta + 2i(\nabla, a) + a^2 + V(h) - [\nabla, a]_z \end{pmatrix}.$$

In this part we shall obtain asymptotic formulas for the eigenvalues of the operator $H(a,V(h))$ by two schemes.

In the first method we divide the operator $H(a,V(h))$ by the following parts:

$$H(a,V(h)) = \begin{pmatrix} -\Delta + 2i(\nabla,a) + a^2 + [\nabla,a]_z & [\nabla,a]_x - i[\nabla,a]_y \\ [\nabla,a]_x + i[\nabla,a]_y & -\Delta + 2i(\nabla,a) + a^2 - [\nabla,a]_z \end{pmatrix} + \begin{pmatrix} V(h) & 0 \\ 0 & V(h) \end{pmatrix} \qquad (2.1.5)$$

Denote by the operator generated in $L_2(F) \times L_2(F)$ by the expression

$$H_0(a) = (-i\nabla - a)^2 \cdot I + \sigma \cdot B \qquad (2.1.6)$$

and boundary conditions (2.1.2).

Let us find eigenvalues and eigenfunctions of the operator $H_0'(a)$. We seek the eigenfunctions in the form

$$\phi(h) = \begin{pmatrix} c_1 e^{i(\gamma+t,h)} \\ c_2 e^{i(\gamma+t,h)} \end{pmatrix} \qquad (2.1.7)$$

Then eigenvalues will satisfy the equation

$$H_0'(a)\phi(h) = \lambda\phi(h)$$

Substituting the expression for $\phi(h)$ from (2.1.7) we get the following system of equations

$$\begin{cases} c_1\left(|\gamma+t|^2 - 2(\gamma+t,a) + a^2 + i[\gamma+t,a]_z\right) + c_2\left(i[\gamma+t,a]_x + [\gamma+t,a]_y\right) = \lambda c_1 \\ c_2\left(|\gamma+t|^2 - 2(\gamma+t,a) + a^2 - i[\gamma+t,a]_z\right) + c_1\left(i[\gamma+t,a]_x - [\gamma+t,a]_y\right) = \lambda c_2 \end{cases} \qquad (2.1.8)$$

In order to (2.1.8) have nontrivial solutions its determinant should be equal to zero

$$\begin{vmatrix} |\gamma+t|^2 - 2(\gamma+t,a) + a^2 + i[\gamma+t,a]_z - \lambda & i[\gamma+t,a]_x + [\gamma+t,a]_y \\ i[\gamma+t,a]_x - [\gamma+t,a]_y & |\gamma+t|^2 - 2(\gamma+t,a) + a^2 - i[\gamma+t,a]_z - \lambda \end{vmatrix} = 0 \qquad (2.1.9)$$

From this we get

$$\lambda^2 - 2\lambda\left(|\gamma+t|^2 - 2(\gamma+t,a) + a^2\right) + \left(|\gamma+t|^2 - 2(\gamma+t,a) + a^2\right)^2 + [\gamma+t,a]_z^2 + [\gamma+t,a]_x^2 + [\gamma+t,a]_y^2 = 0$$

Solving this quadratic equation we obtain

$$\lambda_{\gamma^-} = |\gamma+t|^2 - 2(\gamma+t,a) + a^2 - i\big|[\gamma+t,a]\big|, \qquad (2.1.10)$$

$$\lambda_{\gamma^+} = |\gamma+t|^2 - 2(\gamma+t,a) + a^2 + i\big|[\gamma+t,a]\big|. \qquad (2.1.11)$$

Since as is known from perturbation theory the periodic function may change the eigenvalue of the operator $H_0'(a)$ by the order $O(1)$, from (2.1.5) we get the following:

Theorem 2.1.1. For any $\lambda_\gamma \in \sigma\left(H_0^t(a)\right)$ there exists the corresponding eigenvalue $\Lambda \in \sigma\left(H_t\left(a, V(h)\right)\right)$ that satisfies to the asymptotics

$$\Lambda = |\gamma + t|^2 - 2(\gamma + t, a) + a^2 + i\big[[\gamma + t, a]\big] + O(1) \tag{2.1.12}$$

In the second method we divide the operator $H\left(a, V(h)\right)$ by the following parts

$$H\left(a, V(x)\right) = \begin{pmatrix} -\Delta + 2i(\nabla, a) + a^2 + [\nabla, a]_z & 0 \\ 0 & -\Delta + 2i(\nabla, a) + a^2 - [\nabla, a]_z \end{pmatrix} +$$

$$+ \begin{pmatrix} V(h) & [\nabla, a]_x - i[\nabla, a]_y \\ [\nabla, a]_x + i[\nabla, a]_y & V(h) \end{pmatrix} \tag{2.1.13}$$

Denote by $M_t(a)$ the operator generated in $L_2(F) \times L_2(F)$ by the expression

$$M(a) = \begin{pmatrix} -\Delta + 2i(\nabla, a) + a^2 + [\nabla, a]_z & 0 \\ 0 & -\Delta + 2i(\nabla, a) + a^2 - [\nabla, a]_z \end{pmatrix} \tag{2.1.14}$$

And boundary conditions (2.1.2) and by $N(h)$ the expression

$$N(h) = \begin{pmatrix} V(h) & [\nabla, a]_x - i[\nabla, a]_y \\ [\nabla, a]_x + i[\nabla, a]_y & V(h) \end{pmatrix}. \tag{2.1.15}$$

Writing the eigenvalue problem for the operator $M(a)$ we get

$$\begin{cases} \left(-\Delta + 2i(\nabla, a) + a^2 + [\nabla, a]_z\right)u_1 = \lambda u_1 \\ \left(-\Delta + 2i(\nabla, a) + a^2 - [\nabla, a]_z\right)u_2 = \lambda u_2 \end{cases}, \text{ where } u = \begin{pmatrix} u_1 \\ u_2 \end{pmatrix}.$$

Let

$$\phi_\gamma(h) = \begin{pmatrix} 0 \\ e^{i(\gamma + t, h)} \end{pmatrix}, \qquad \text{where } \gamma \in \Gamma, \ t \in F. \tag{2.1.16}$$

Then

$$\lambda_\gamma = |\gamma + t|^2 - 2(\gamma + t, a) + a^2 - i[\gamma + t, a]_z. \tag{2.1.17}$$

Similarly, if

$$\phi^\gamma(h) = \begin{pmatrix} e^{i(\gamma + t, h)} \\ 0 \end{pmatrix}, \qquad \text{where } \gamma \in \Gamma, \ t \in F, \tag{2.1.18}$$

тогда

$$\lambda^\gamma = |\gamma + t|^2 - 2(\gamma + t, a) + a^2 + i[\gamma + t, a]_z. \tag{2.1.19}$$

It is also possible to prove that the other eigenvalues of the operator $M_t(a)$ are absent. In other words, the spectrum of the operator $M_t(a)$ consists of λ_γ and λ^γ.

Denote by Λ_N and Φ_N eigenvalues and eigenfunctios of the operator $H_t(a,V(h))$. Let the index N is such that

$$\left|\left(\Phi_N,\phi_\gamma\right)\right|=\max_\beta\left\{\left|\left(\Phi_N,\phi_\beta\right)\right|,\left|\left(\Phi_N,\phi^\beta\right)\right|\right\}.$$

The corresponding to this index eigenvalue Λ_N we denote as $\Lambda_N(\lambda_\gamma)$.

We use the known equality

$$\left(\Lambda_N-\lambda_\gamma\right)\left(\Phi_N,\phi_\gamma\right)=\left(\Phi_N,N(h)\phi_\gamma\right).\tag{2.1.20}$$

For the sake of simplicity assume that the potential $V(h)$ is a trigonometric polynomial

$$V(h)=\sum_{\gamma\in Q}V_\gamma e^{i(\gamma,h)},\quad\text{where}\quad V_\gamma=\int_F V(h)e^{-i(\gamma,h)}dh,\quad Q=\left\{\gamma\in\Gamma:V_\gamma\neq0\right\}.$$

Since the system of eigenfunctions (2.1.16) and (2.1.18) of the operator $M_t(a)$ form full system in $L_2(F)$, $N(h)\phi_\gamma$ may be decomposed by them

$$N(h)\phi_\gamma=\sum_\beta\left(N(h)\phi_\gamma,\phi_\beta\right)\phi_\beta+\sum_\beta\left(N(h)\phi_\gamma,\phi^\beta\right)\phi^\beta,\tag{2.1.21}$$

where $\left(N(h)\phi_\gamma,\phi_\beta\right)=\begin{cases}0,&\beta\neq\gamma\\ \left[\gamma+t,a\right]_x+i\left[\gamma+t,a\right]_y,&\beta=\gamma\end{cases}$

$$\left(N(h)\phi_\gamma,\phi_{\gamma+\gamma_1}\right)=\left(N(h)\phi^\gamma,\phi^{\gamma+\gamma_1}\right)=V_{\gamma_1}$$

$$\left(N(h)\phi^\gamma,\phi^\beta\right)=\begin{cases}0,&\beta\neq\gamma\\ \left[\gamma+t,a\right]_x-i\left[\gamma+t,a\right]_y,&\beta=\gamma\end{cases}.$$

Denote:

$$c^\gamma=\left[\gamma+t,a\right]_x-i\left[\gamma+t,a\right]_y,\tag{2.1.22}$$

$$b^\gamma=\left[\gamma+t,a\right]_x+i\left[\gamma+t,a\right]_y.\tag{2.1.23}$$

Now from (2.1.20) we have

$$\left(\Lambda_N-\lambda_\gamma\right)\left(\Phi_N,\phi_\gamma\right)=\sum_{\gamma_1}V_{\gamma_1}\left(\Phi_N,\phi_{\gamma+\gamma_1}\right)+c^\gamma\left(\Phi_N,\phi^\gamma\right).\tag{2.1.24}$$

From this follows that

$$\left|\Lambda_N-\lambda_\gamma\right|<c^\gamma+O(1).\tag{2.1.25}$$

Allying (2.1.20) in (2.1.24) we obtain

$$\left(\Lambda_N - \lambda_\gamma\right)\left(\Phi_N, \phi_\gamma\right) = O(1)\left(\Phi_N, \phi_\gamma\right) + \frac{c^\lambda \sum_{\gamma_1} V_{\gamma_1}\left(\Phi_N, \phi^{\gamma+\gamma_1}\right)}{\Lambda_N - \lambda^\gamma} + \frac{c^\gamma b^\gamma}{\Lambda_N - \lambda^\gamma}\left(\Phi_N, \phi_\gamma\right) \tag{2.1.26}$$

Now let us analyze the expression

$$\frac{c^\lambda \sum_{\gamma_1} V_{\gamma_1}\left(\Phi_N, \phi^{\gamma+\gamma_1}\right)}{\Lambda_N - \lambda^\gamma} \tag{2.1.27}$$

Assume that

$$\left|\lambda^\gamma - \lambda_\gamma\right| > 2c^\gamma \tag{2.1.28}$$

Then from the formulas (2.1.25) и (2.1.28) follows that

$$\left|\Lambda_N - \lambda^\gamma\right| = \left|\Lambda_N - \lambda^\gamma - \lambda_\gamma + \lambda_\gamma\right| > \left|\lambda^\gamma - \lambda_\gamma\right| - \left|\Lambda_N - \lambda_\gamma\right| > 2c^\gamma - c^\gamma = c^\gamma \tag{2.1.29}$$

The last formula gives

$$\frac{c^\lambda \sum_{\gamma_1} V_{\gamma_1}\left(\Phi_N, \phi^{\gamma+\gamma_1}\right)}{\Lambda_N - \lambda^\gamma} = O(1)\left(\Phi_N, \phi_\gamma\right) \tag{2.1.30}$$

Considering this in (2.1.24) we obtain

$$\Lambda_N - \lambda_\gamma - \frac{c^\gamma b^\gamma}{\Lambda_N - \lambda^\gamma} = O(1)$$

Substituting $\lambda^\gamma = \lambda_\gamma + 2[\gamma + t, a]_z$ instead of λ^γ, we have

$$\Lambda_N - \lambda_\gamma - \frac{c^\gamma b^\gamma}{\Lambda_N - \lambda_\gamma - 2[\gamma + t, a]_z} = O(1)$$

If to solve this equation with respect to $\Lambda_N - \lambda_\gamma$ we arrive too the following quadratic equation

$$\left(\Lambda_N - \lambda_\gamma\right)^2 - \left(2[\gamma + t, a]_z + O(1)\right)\left(\Lambda_N - \lambda_\gamma\right) - c^\gamma b^\gamma + 2[\gamma + t, a]_z \cdot O(1) = 0$$

$$\Lambda_N - \lambda_\gamma = [\gamma + t, a]_z + O(1) \mp \sqrt{\left([\gamma + t, a]_z + O(1)\right)^2 + c^\gamma b^\gamma - 2[\gamma + t, a]_z \cdot O(1)} =$$

$$= [\gamma + t, a]_z + O(1) \mp \sqrt{[\gamma + t, a]_z^2 + c^\gamma b^\gamma + O(1)}.$$

Since c^γ, b^γ и $[\gamma + t, a]_z \approx |\gamma|$, we have

$$\Lambda_N = \lambda_\gamma + [\gamma + t, a]_z \mp \sqrt{[\gamma + t, a]_z^2 + c^\gamma b^\gamma} + O(1) \tag{2.1.31}$$

Considering $c^\gamma b^\gamma = [\gamma + t, a]_x^2 + [\gamma + t, a]_y^2$ (see (2.1.22) and (2.1.23)), (2.1.31) may be written as

$$\Lambda_N = \lambda_\gamma + [\gamma + t, a]_z \mp \left|[\gamma + t, a]\right| + O(1) \tag{2.1.32}$$

Now we analyze the condition (2.1.28). Taking into account the formulas (2.1.17), (2.1.19) and(2.1.22) we obtain

$$2\left|\left[\gamma+t,a\right]_z\right|>2\left|\left[\gamma+t,a\right]_x-i\left[\gamma+t,a\right]_y\right|$$

From this

$$\left|\left[\gamma+t,a\right]_z\right|>\left|\left[\gamma+t,a\right]_x\right|+\left|\left[\gamma+t,a\right]_y\right| \tag{2.1.33}$$

Similarly one can get $|v_z|>|v_x|+|v_y|$. This condition geometrically means that the vector v lies near

the axis OZ or $\left(\vec{v}^\wedge\overline{OZ}\right)<\dfrac{\pi}{4}$.

Theorem 2.1.2. If $\left|\left[\gamma+t,a\right]_z\right|>\left|\left[\gamma+t,a\right]_x\right|+\left|\left[\gamma+t,a\right]_y\right|$, then for any λ_γ the corresponding

eigenvalue $\Lambda_N\left(\lambda_\gamma\right)$ of the operator $H_t\left(a,V\left(h\right)\right)$ satisfies to the asymptotics (2.1.31) or (2.1.32).

Following $\left[6\right]$ the theorem below also may be proved.

Theorem 2.1.3. If $\left|\left[\gamma+t,a\right]_z\right|>\left|\left[\gamma+t,a\right]_x\right|+\left|\left[\gamma+t,a\right]_y\right|$, then for any λ_γ there exists the

corresponding eigenvalue Λ_N of the operator $H_t\left(a,V\left(h\right)\right)$ satisfying the asymptotics (2.1.31) or

(2.1.32).

2.2. Asymptotic formulas for a series of non-resonant
eigenvalues of three-dimensional periodic Pauli operator

In this chapter we consider operator $H_t\left(a,V\left(h\right)\right)$ generated in $L_2\left(F\right)\times L_2\left(F\right)$ by the

expression

$$H\left(a,V\left(h\right)\right)=\left(\left(-i\nabla-a\right)^2+V\left(h\right)\right)\cdot I+\sigma\cdot B \tag{2.1.1}$$

and boundary conditions

$$u\left(h+\omega_j\right)=e^{2\pi i t_j}u\left(h\right) \qquad j=1,3 , \tag{2.1.2}$$

where $h=\left(x,y,z\right)\in R^3$, $B=\left[\nabla,a\right]$ is a magnet field generated by the vector potential

$a=\left(a_1,a_2,a_3\right)$, $B=\left(B_1,B_2,B_3\right)$

$$I=\begin{pmatrix}1 & 0\\ 0 & 1\end{pmatrix}, \quad B_1=\begin{vmatrix}\dfrac{\partial}{\partial y} & \dfrac{\partial}{\partial z}\\ a_2 & a_3\end{vmatrix}, \quad B_2=\begin{vmatrix}\dfrac{\partial}{\partial z} & \dfrac{\partial}{\partial x}\\ a_3 & a_1\end{vmatrix}, \quad B_3=\begin{vmatrix}\dfrac{\partial}{\partial x} & \dfrac{\partial}{\partial y}\\ a_1 & a_2\end{vmatrix}, \tag{2.1.3}$$

$$\sigma_1=\begin{pmatrix}0 & 1\\ 1 & 0\end{pmatrix}, \quad \sigma_2=\begin{pmatrix}0 & -i\\ i & 0\end{pmatrix}, \quad \sigma_3=\begin{pmatrix}1 & 0\\ 0 & -1\end{pmatrix}, \tag{2.1.4}$$

F is a fundamental domain for some lattice $\Omega = \{m_1\omega_1 + m_2\omega_2 + m_3\omega_3 : m_1, m_2, m_3 \in Z\}$, i.e. parallelepiped, $t = t_1\gamma^1 + t_2\gamma^2 + t_3\gamma^3$, $\gamma^1, \gamma^2, \gamma^3$ are biorthogonal to $\omega_1, \omega_2, \omega_3$ vectors, F^*-fundamental domain of the lattice $\Gamma = \{n_1\gamma^1 + n_2\gamma^2 + n_3\gamma^3 : n_1, n_2, n_3 \in Z\}$, $V(h)$ is periodic, smooth enough function.

Considering (2.1.3) and (2.1.4) the operator $H\big(a, V(h)\big)$ may written in the following explicit form

$$H\big(a, V(h)\big) = \begin{pmatrix} -\Delta + 2i(\nabla, a) + a^2 + V(h) + [\nabla, a]_z & [\nabla, a]_x - i[\nabla, a]_y \\ [\nabla, a]_x + i[\nabla, a]_y & -\Delta + 2i(\nabla, a) + a^2 + V(h) - [\nabla, a]_z \end{pmatrix}.$$

In this section we obtain higher-order asymptotic formulas for the eigenvalues of Pauli operator $H\big(a, V(h)\big)$ as in paragraph 1 by two methods. We will divide the Pauli operator into two parts, where the first part we consider as the unperturbed operator and the second part as a perturbation.

In the first method the operator $H(a, V(h))$ is divided as follows

$$H(a, V(h)) = \begin{pmatrix} -\Delta + 2i(\nabla, a) + a^2 + [\nabla, a]_z & [\nabla, a]_x - i[\nabla, a]_y \\ [\nabla, a]_x + i[\nabla, a]_y & -\Delta + 2i(\nabla, a) + a^2 - [\nabla, a]_z \end{pmatrix} + \begin{pmatrix} V(h) & 0 \\ 0 & V(h) \end{pmatrix}. \tag{2.1.5}$$

Denote by $H_0^t(a)$ the operator generated in $L_2(F) \times L_2(F)$ by expression

$$H_0(a) = (-i\nabla - a)^2 \cdot I + \sigma \cdot B \tag{2.1.6}$$

And boundary conditions (2.1.2).

Then the eigenvalues of the operator $H_0^t(a)$ will be the following:

$$\lambda_{\gamma^-} = |\gamma + t|^2 - 2(\gamma + t, a) + a^2 - i\big|[\gamma + t, a]\big|, \tag{2.1.10}$$

$$\lambda_{\gamma^+} = |\gamma + t|^2 - 2(\gamma + t, a) + a^2 + i\big|[\gamma + t, a]\big|. \tag{2.1.11}$$

Now, using these formulas we find eigenfunctions. It follows from (2.1.10), (2.1.11) and (2.1.8) that

$$\begin{cases} c_1\big(i[\gamma + t, a]_z + \big|[\gamma + t, a]\big|\big) + c_2\big(i[\gamma = t, a]_x + [\gamma + t, a]_y\big) = 0 \\ c_1\big(i[\gamma + t, a]_x - [\gamma = t, a]_y\big) + c_2\big(-i[\gamma = t, a]_z + \big|[\gamma + t, a]\big|\big) = 0 \end{cases}$$

From this we get

$$c_1 = \frac{i[\gamma + t, a]_z - \big|[\gamma + t, a]\big|}{-i[\gamma + t, a]_x + [\gamma + t, a]_y} c_2 \tag{2.2.1}$$

Now we choose c_1, c_2 such that $\phi(h)$ be normalized i.e. $c_1^2 + c_2^2 = 1$. The use of (2.2.1) gives

$$\left(1+\frac{i[\gamma+t,a]_z-[\![\gamma+t,a]\!]}{-i[\gamma+t,a]_x+[\gamma+t,a]_y}\right)c_2{}^2=1.$$

Solving this equation we obtain

$$\frac{-[\gamma+t,a]_z{}^2-2i[\gamma+t,a]_z\cdot[\gamma+t,a]+[\gamma+t,a]^2-[\gamma+t,a]_x{}^2}{-[\gamma+t,a]_x{}^2-2i[\gamma+t,a]_x[\gamma+t,a]_y+[\gamma+t,a]_y{}^2}+$$

$$+\frac{[\gamma+t,a]_y{}^2-2i[\gamma+t,a]_x\cdot[\gamma+t,a]_y}{-[\gamma+t,a]_x{}^2-2i[\gamma+t,a]_x[\gamma+t,a]_y+[\gamma+t,a]_y{}^2}=\frac{1}{c_2{}^2}.$$

From this we

$$c_2=\pm\frac{i[\gamma+t,a]_z-[\gamma+t,a]_y}{\sqrt{2[\gamma+t,a]_y{}^2-2i[\gamma+t,a]_z\cdot[\gamma+t,a]-2i[\gamma+t,a]_x\cdot[\gamma+t,a]_y}}. \qquad (2.2.2)$$

If to consider (2.2.1) we get

$$c_1=\pm\frac{i[\gamma+t,a]_z-[\![\gamma+t,a]\!]}{\sqrt{2[\gamma+t,a]_y{}^2-2i[\gamma+t,a]_z\cdot[\gamma+t,a]-2i[\gamma+t,a]_x\cdot[\gamma+t,a]_y}}. \qquad (2.2.3)$$

Now if take into account (2.2.2.) and (2.2.3) in (2.1.7) we get that eigenfunction of the operator $H_0^t(a)$ indeed are

$$\phi_{\gamma^+}(h)=\begin{pmatrix}\dfrac{i[\gamma+t,a]_x-[\gamma+t,a]_y}{A}\cdot e^{i(\gamma+t,x)}\\[2mm]\dfrac{i[\gamma+t,a]_z-[\![\gamma+t,a]\!]}{A}\cdot e^{i(\gamma+t,x)}\end{pmatrix},$$

$$\phi_{\gamma^-}(h)=\begin{pmatrix}-\dfrac{i[\gamma+t,a]_x-[\gamma+t,a]_y}{A}\cdot e^{i(\gamma+t,x)}\\[2mm]-\dfrac{i[\gamma+t,a]_z-[\![\gamma+t,a]\!]}{A}\cdot e^{i(\gamma+t,x)}\end{pmatrix},$$

where

$$A=\sqrt{2[\gamma+t,a]_y{}^2-2i[\gamma+t,a]_z\cdot[\gamma+t,a]-2i[\gamma+t,a]_x\cdot[\gamma+t,a]_y}.$$

For the sake of simplicity we assume that the potential $V(h)$ is a trigonometric polynomial

$$V(h)=\sum_{\gamma\in Q}V_\gamma e^{i(\gamma,h)}, \quad \text{где} \quad V_\gamma=\int_F V(h)e^{-i(\gamma,h)}dh, \quad Q=\{\gamma\in\Gamma:V_\gamma\neq0\}.$$

Then

$$\left(\begin{pmatrix}V(h)&0\\0&V(h)\end{pmatrix}\phi_{\gamma^+}(h),\phi_{\beta^-}(h)\right)=\left(\begin{pmatrix}\dfrac{B}{A}\sum_{\gamma_1\in Q}V_{\gamma_1}e^{i(\gamma+\gamma_1+t,h)}\\[2mm]\dfrac{C}{A}\sum_{\gamma_1\in Q}V_{\gamma_1}e^{i(\gamma+\gamma_1+t,h)}\end{pmatrix},\begin{pmatrix}\dfrac{B}{A}e^{i(\beta+t,h)}\\[2mm]\dfrac{C}{A}e^{i(\beta+t,h)}\end{pmatrix}\right)=$$

$$= \begin{cases} 0, & \beta \neq \gamma + \gamma_1 \\ \dfrac{B^2 + C^2}{A^2} \displaystyle\sum_{\gamma_1 \in Q} V_{\gamma_1}, & \beta = \gamma + \gamma_1 \end{cases}.$$

Here

$$B = i[\gamma + t, a]_x - [\gamma + t, a]_y, \quad C = i[\gamma + t, a]_z - [[\gamma + t, a]]$$

Since

$$B^2 + C^2 = \left(i[\gamma + t, a]_x - [\gamma + t, a]_y\right)^2 + \left(i[\gamma + t, a]_z - [[\gamma + t, a]]\right)^2 =$$

$$= 2[\gamma + t, a]_y^2 - 2i[\gamma + t, a]_z \cdot [\gamma + t, a] - 2i[\gamma + t, a]_x \cdot [\gamma + t, a]_y = A^2$$

Thus

$$\left(V(h) \cdot I \cdot \phi_{\gamma^-}(h), \phi_{\beta^+}(h)\right) = \begin{cases} 0, & \beta \neq \gamma + \gamma_1 \\ \displaystyle\sum_{\gamma_1 \in Q} V_{\gamma_1}, & \beta = \gamma + \gamma_1 \end{cases}$$

By the definition we get that λ_{γ^+} is associated with $\lambda_{(\gamma + \gamma_1)^+}$ by $\gamma_1 \in Q$ ($\gamma_1 \in Q^{(m)}$) through the perturbation $V(h)$ ($V^{(m)}(h)$). By the same way may be shown that λ_{γ^-} is associated with $\lambda_{(\gamma + \gamma_1)^-}$ through the perturbation $V(h)$ ($V^{(m)}(h)$).

Similarly we show that

$$\left(V(h) \cdot I \cdot \varphi_{\gamma^+}(h), \varphi_{\beta^-}(h)\right) = \begin{cases} 0, & \beta \neq \gamma + \gamma_1 \\ -\displaystyle\sum_{\gamma_1 \in Q} V_{\gamma_1}, & \beta = \gamma + \gamma_1 \end{cases}.$$

Thus λ_{γ^+} associated with $\lambda_{(\gamma + \gamma_1)^-}$ by $\gamma_1 \in Q$ ($\gamma_1 \in Q^{(m)}$) through the perturbation $V(h)$ ($V^{(m)}(h)$). By the same way may be shown that λ_{γ^-} is associated with $\lambda_{(\gamma + \gamma_1)^+}$ through the perturbation $V(h)$ ($V^{(m)}(h)$).

Now let us estimate the difference

$$\left|\lambda_{\gamma^+} - \lambda_{(\gamma + \gamma_1)^+}\right| = \left|\left|\gamma + t\right|^2 - 2(\gamma + t, a) + a^2 + [\gamma + t, a] - \left|\gamma + \gamma_1 + t\right|^2 + 2(\gamma + \gamma_1 + t, a) - a^2 - [\gamma + \gamma_1 + t, a]\right| \geq$$

$$\geq \left|\left|\gamma + t\right|^2 - \left|\gamma + \gamma_1 + t\right|^2\right| - \left|2(\gamma_1, a) - [\gamma_1, a]\right|.$$

Let γ satisfies

$$\left|\left|\gamma + t\right|^2 - \left|\gamma + \gamma_1 + t\right|^2\right| > \left|\gamma + t\right|^\alpha.$$

One easily may show that

$$\left|2(\gamma_1, a) - [\gamma_1, a]\right| < C|\gamma_1||a| < M|\gamma + t| < \frac{1}{2}|\gamma + t|^\alpha$$

From this we get that

$$\left|\lambda_{\gamma^+} - \lambda_{(\gamma+\gamma_1)^+}\right| > \frac{1}{2}|\gamma+t|^\alpha$$

By the same way may be proved that

$$\left|\lambda_{\gamma^-} - \lambda_{(\gamma+\gamma_1)^-}\right| > \frac{1}{2}|\gamma+t|^\alpha$$

Let us estimate $\left|\lambda_{\gamma^+} - \lambda_{(\gamma+\gamma_1)^-}\right|$, for $\gamma_1 \in Q$

$$\left|\lambda_{\gamma^+} - \lambda_{(\gamma+\gamma_1)^-}\right| = \left||\gamma+t|^2 - 2(\gamma+t,a) + a^2 + [\gamma+t,a] - |\gamma+\gamma_1+t|^2 + 2(\gamma+\gamma_1+t,a) - a^2 + [\gamma+\gamma_1+t,a]\right| \geq$$

$$\geq \left||\gamma+t|^2 - |\gamma+\gamma_1+t|^2\right| - |2(\gamma_1,a) - [\gamma_1,a]| - 2[\gamma+t,a] > \frac{1}{2}|\gamma+t|^\alpha - 2[\gamma+t,a]$$

Let the following condition be fulfilled

$$[\gamma+t,a] < \frac{1}{5}|\gamma+t|^\alpha. \qquad (2.2.4)$$

Then

$$\left|\lambda_{\gamma^+} - \lambda_{(\gamma+\gamma_1)^-}\right| > \frac{1}{10}|\gamma+t|^\alpha.$$

Thus if $\gamma+t \notin V_{\gamma_1}(\alpha)$ and (2.2.4) is fulfilled then the m-order non-resonance condition for the eigenvalue λ_{γ^+} takes the form

$$\left|\lambda_{\gamma^+} - \lambda_{(\gamma+\gamma_1)^-}\right| > \frac{1}{10}|\gamma+t|^\alpha \text{ and } \left|\lambda_{\gamma^+} - \lambda_{(\gamma+\gamma_1)^+}\right| > \frac{1}{2}|\gamma+t|^\alpha,$$

$$V_\gamma(\alpha) = \left\{h \in R^3 : \left||h|^2 - |h+\gamma|^2\right| < |h|^\alpha\right\}, \qquad V^{(m)}(\alpha) = \bigcup_{\gamma_1 \in Q^{(m)}} V_{\gamma_1}(\alpha), \qquad \overline{V^{(m)}}(\alpha) = R^3 \setminus V^{(m)}(\alpha).$$

Note that the number of the vectors γ involved in $Q^{(m)}$ is not greater than $|Q|^m$, i.e. $\left|Q^{(m)}\right| \leq |Q|^m$, where $|Q|$ is a number of the vectors involved in Q. Therefore if $\rho \to \infty$, then

$$\mu\left(V_{\gamma_1}(\alpha) \cap S_\rho\right) = \mu(S_\rho) \cdot O(\rho^{\alpha-1}); \qquad \mu\left(\overline{V^{(m)}}(\alpha) \cap S_\rho\right) = \mu(S_\rho) \cdot \left\{1 + O(\rho^{\alpha-1})\right\},$$

where $S_\rho = \left\{h \in R^3 : |h| = \rho\right\}$.

Thus $\overline{V^{(m)}}(\alpha)$ constitutes the main part of R^3. This means that m-order non-resonance eigenvalues constitute the bulk of all eigenvalues. If the condition $\sum_{\gamma \in Q}|V_\gamma| < \infty$ is true then the following theorems are valid.

Theorem 2.2.1. a) If the eigenvalue λ_{γ^+} is m-order nion0-resonance i.e. . $\gamma+t \in \overline{V^{(m)}}(\alpha)$ and (2.2.4) is valid than corresponding eigenvalue $\Lambda_N\left(\lambda_{\gamma^+}\right)$ of the operator $H_t(a, V(h))$ satisfies to the formula

$$\Lambda_N\left(\lambda_{\gamma^*}\right)=\lambda_{\gamma^*}+\overline{F_{m-1}}\left(\lambda_{\gamma^*},\lambda_{\gamma^-},V(h)\right)+O\left(\left|\gamma+t\right|^{-m\alpha}\right),\tag{2.2.5}$$

where $\overline{F_k}\left(\lambda_{\gamma^*},\lambda_{\gamma^-},V(h)\right)$ is explicitly expressed by λ_{γ^*}, λ_{γ^-}, $V(h)$ and has an order $O\left(\left|\gamma+t\right|^{-k\alpha}\right)$.

б) For any m-order non-resonance eigenvalue λ_{γ^*} there exists an eigenvalue Λ_N of the operator $H_t\left(a,V(h)\right)$ satisfying to (2.2.5).

Theorem 2.2.2. a) Let the eigenvalue $\Lambda_N(t)$ of the operator $H_t\left(a,V(h)\right)$ be such that

$$\left|\Lambda_N-\lambda_{\gamma^*}\right|>2M \qquad\text{and}\qquad \left|\left(\Psi_{N,t}(h),\phi_{\gamma^*}(h)\right)\right|>\left|\gamma+t\right|^{-m_0\alpha},$$

where $M\geq\sum_{\gamma}V_{\gamma}\geq\sup_h V(h)$.

If λ_{γ^*} is m-order non-resonance, then eigenvalue $\Lambda_N\left(\lambda_{\gamma^*}\right)$ satisfies the asymptotics

$$\Lambda_N\left(\lambda_{\gamma^*}\right)=\lambda_{\gamma^*}+\overline{F_{m-m_0-1}}\left(\lambda_{\gamma^*},\lambda_{\gamma^-},V(h)\right)+O\left(\left|\gamma+t\right|^{-(m-m_0)\alpha}\right).\tag{2.2.6}$$

б) Eigenvalue $\Lambda_{k\lambda}$ for which is valid

$$\left|\left(\Psi_{k,t}(h),\phi_{\gamma^*}(h)\right)\right|=\max_N\left\{\left|\left(\Psi_{N,t}(h),\phi_{\gamma^*}(h)\right)\right|,\left|\left(\Psi_{N,t}(h),\phi_{\gamma^-}(h)\right)\right|\right\}$$

satisfies the asymptotics (2.2.6) с $m_0=\left[\frac{n}{2\alpha}\right]$, $k\equiv k\lambda$.

In the second method we divide the operator $H\left(a,V(h)\right)$ into the following parts

$$H\left(a,V(x)\right)=\begin{pmatrix}-\Delta+2i(\nabla,a)+a^2+[\nabla,a]_z & 0\\ 0 & -\Delta+2i(\nabla,a)+a^2-[\nabla,a]_z\end{pmatrix}+$$

$$+\begin{pmatrix}V(h) & [\nabla,a]_x-i[\nabla,a]_y\\ [\nabla,a]_x+i[\nabla,a]_y & V(h)\end{pmatrix}$$

This scheme is more simple and applicable to the $a\equiv a(h)$.

Denote by $M_t(a)$ the operator generated in $L_2(F)\times L_2(F)$ by the expression

$$M(a)=\begin{pmatrix}-\Delta+2i(\nabla,a)+a^2+[\nabla,a]_z & 0\\ 0 & -\Delta+2i(\nabla,a)+a^2-[\nabla,a]_z\end{pmatrix}.\tag{2.2.7}$$

And boundary conditions (2.1.2) and by $N(h)$ the following expression

$$N(h)=\begin{pmatrix}V(h) & [\nabla,a]_x-i[\nabla,a]_y\\ [\nabla,a]_x+i[\nabla,a]_y & V(h)\end{pmatrix}.\tag{2.2.8}$$

In the first chapter was proved that eigenvalues and eigenfunctions of the operator $M_t(a)$ indeed are

$$\lambda_\gamma = |\gamma+t|^2 - 2(\gamma+t,a) + a^2 - i[\gamma+t,a]_z,$$ (2.2.9)

$$\lambda^\gamma = |\gamma+t|^2 - 2(\gamma+t,a) + a^2 + i[\gamma+t,a]_z;$$ (2.2.10)

$$\phi_\gamma(h) = \begin{pmatrix} 0 \\ e^{i(\gamma+t,h)} \end{pmatrix}, \quad \text{where} \quad \gamma \in \Gamma, \ t \in F,$$ (2.2.11)

$$\phi^\gamma(h) = \begin{pmatrix} e^{i(\gamma+t,h)} \\ 0 \end{pmatrix}, \quad \text{where} \quad \gamma \in \Gamma, \ t \in F.$$ (2.2.12)

So, the spectrum $M_t(a)$ consists of λ_γ and λ^γ, where $\gamma \in \Gamma$. Now we find the asymptotic formulas for the eigenvalues of the operator $H_t(a, V(h))$. For the sake of simplicity we assume that the potential $V(h)$ is a trigonometric polynomial

$$V(x) = \sum_{\gamma \in Q} V_\gamma e^{i(\gamma,h)} \ , \text{where где} \ V_\gamma = \int_F V(h) e^{-i(\gamma,h)} dh, \ Q = \{\gamma \in \Gamma : V_\gamma \neq 0\}.$$

Moreover, without loss of generality, we assume

$$V_0 = \int_F V(h) dh = 0.$$ (2.2.13)

In solving this method we impost the following conditions:

    I. $a \in XOY$, i.e. $a = (a_1, a_2, 0)$;

    II. $v = (0,0,1) \notin Q^{(m)}$;

    III. $[\gamma+t,a]_z < c \cdot |\gamma+t|^{\alpha^1}$.

First, we find the formulas that will continue to use

$$N(h) = \begin{pmatrix} V(h) & [\nabla,a]_x - i[\nabla,a]_y \\ [\nabla,a]_x + i[\nabla,a]_y & V(h) \end{pmatrix} = \begin{pmatrix} V(h) & (a_2 + ia_1)\dfrac{\partial}{\partial z} - \left(i\dfrac{\partial}{\partial x} + \dfrac{\partial}{\partial y}\right)a_3 \\ (a_2 - ia_1)\dfrac{\partial}{\partial z} + \left(i\dfrac{\partial}{\partial x} + \dfrac{\partial}{\partial y}\right)a_3 & V(h) \end{pmatrix}.$$

Since $a \in XOY$

$$N(h) = \begin{pmatrix} V(h) & (a_2 + ia_1)\dfrac{\partial}{\partial z} \\ (a_2 - ia_1)\dfrac{\partial}{\partial z} & V(h) \end{pmatrix}.$$ (2.2.14)

Now we find the eigenvalues associated with λ_γ through $N(h)$. For this purpose we calculate $N(h)\phi_\gamma(h)$ and $N(h)\phi^\gamma(h)$.

$$N(h)\phi_\gamma(h) = \begin{pmatrix} (a_2+ia_1)\cdot(\gamma+t)_z \cdot e^{i(\gamma+t,h)} \\ \sum_{\gamma_1\in Q} V_{\gamma_1} e^{i(\gamma+\gamma_1+t,h)} \end{pmatrix}, \qquad N(h)\phi^\gamma(h) = \begin{pmatrix} \sum_{\gamma_1\in Q} V_{\gamma_1} e^{i(\gamma+\gamma_1+t,h)} \\ (a_2-ia_1)\cdot(\gamma+t)_z \cdot e^{i(\gamma+t,h)} \end{pmatrix}.$$

From this we obtain that

$$\left\| N(h)\phi_\gamma(h) \right\| < c\left| (\gamma+t)_z \right| \quad \text{и} \quad \left\| N(h)\phi^\gamma(h) \right\| < c\left| (\gamma+t)_z \right|; \tag{2.2.15}$$

$$\left(N(h)\cdot\phi_\gamma(h), \phi_\beta(h) \right) = \begin{cases} 0 & ,\beta\neq\gamma+\gamma_1 \\ \sum_{\gamma_1\in Q} V_{\gamma_1} & ,\beta=\gamma+\gamma_1 \end{cases}; \tag{2.2.16}$$

$$\left(N(h)\cdot\phi^\gamma(h), \phi^\beta(h) \right) = \begin{cases} 0 & ,\beta\neq\gamma+\gamma_1 \\ \sum_{\gamma_1\in Q} V_{\gamma_1} & ,\beta=\gamma+\gamma_1 \end{cases}. \tag{2.2.17}$$

These formulas show that λ_γ is associate with $\lambda_{\gamma+\gamma_1}$, and λ^γ - with c $\lambda^{\gamma+\gamma_1}$ through $N(h)$.

$$\left(N(h)\cdot\phi_\gamma(h), \phi^\beta(h) \right) = \begin{cases} 0 & ,\beta\neq\gamma \\ (a_2+ia_1)\cdot(\gamma+t)_z & ,\beta=\gamma \end{cases}, \tag{2.2.18}$$

$$\left(N(h)\cdot\phi^\gamma(h), \phi^\beta(h) \right) = \begin{cases} 0 & ,\beta\neq\gamma \\ (a_2-ia_1)\cdot(\gamma+t)_z & ,\beta=\gamma \end{cases}. \tag{2.2.19}$$

Denote

$$c^\gamma = (a_2+ia_1)(\gamma+t)_z, \qquad b^\gamma = (a_2-ia_1)(\gamma+t)_z. \tag{2.2.20}$$

As follows from the condition III

$$c^\gamma = O\left(|\gamma+t|^{\alpha^1} \right), \qquad b^\gamma = O\left(|\gamma+t|^{\alpha^1} \right). \tag{2.2.20a}$$

Suppose that $\gamma+t\in \overline{V^{(m)}}(\alpha)$. Then for any $\gamma_1\neq v\in Q^{(m)}$ (see condition II) takes place the relation

$$\left| \lambda_\gamma - \lambda_{\gamma+\gamma_1} \right| = \left| |\gamma+t|^2 - 2(\gamma+t,a) + a^2 + [\gamma+t,a]_z - |\gamma+\gamma_1+t|^2 - 2(\gamma+\gamma_1+t,a) - a^2 + [\gamma+\gamma_1+t,a]_z \right| \geq$$

$$\geq \left| |\gamma+t|^2 - |\gamma+\gamma_1+t|^2 \right| - \left| 2(\gamma_1,a) + [\gamma_1,a]_z \right|.$$

The condition $\gamma+t\in \overline{V^{(m)}}(\alpha)$ means that $\gamma+t\notin V_{\gamma_1}(\alpha)$, i.e.

$$\left| |\gamma+t|^2 - |\gamma+\gamma_1+t|^2 \right| > |\gamma+t|^\alpha, \text{ где } 0<\alpha<1.$$

By easily calculation one can obtain that $\left| 2(\gamma_1,a) + [\gamma_1,a]_z \right| < C|\gamma_1|\cdot|a| < M\cdot|\gamma+t| < \frac{1}{2}|\gamma+t|^\alpha.$

From these inequalities follows that

$$\left| \lambda_\gamma - \lambda_{\gamma+\gamma_1} \right| > \frac{1}{2}|\gamma+t|^\alpha, \quad 0<\alpha^1<\alpha<1. \tag{2.2.21}$$

Similarly one my show that

$$\left|\lambda^\gamma - \lambda^{\gamma+\gamma_1}\right| > \frac{1}{2}\left|\gamma + t\right|^\alpha .$$ (2.2.21a)

The main role in proof of asymptotical formulas plays this known formula

$$\left(\Lambda_N - \lambda_\gamma\right)\left(\Phi_N, \phi_\gamma\right) = \left(\Phi_N, N(h)\phi_\gamma\right).$$ (2.2.23)

Using the decomposition

$$N(h)\phi_\gamma(h) = \sum_\beta \left(N(h)\phi_\gamma, \phi_\beta\right)\phi_\beta(h) + \sum_\beta \left(N(h)\phi_\gamma, \phi^\beta\right)\phi^\beta(h)$$

And considering (2.2.16) - (2.2.19) we get

$$N(h)\phi_\gamma(h) = \sum_{\gamma_1} V_{\gamma_1-\gamma}\phi_{\gamma_1}(h) + c^\gamma \phi^\gamma(h),$$ (2.2.24)

$$N(h)\phi^\gamma(h) = \sum_{\gamma_1} V_{\gamma_1-\gamma}\phi^{\gamma_1}(h) + b^\gamma \phi_\gamma(h).$$ (2.2.25)

Then as follows from (2.2.23)

$$\left(\Lambda_N - \lambda_\gamma\right)\left(\Phi_N, \phi_\gamma\right) = \sum_{\gamma_1} V_{\gamma_1-\gamma}\left(\Phi_N(h), \phi_{\gamma_1}(h)\right) + c^\gamma\left(\Phi_N(h), \phi^\gamma(h)\right).$$ (2.2.26)

From (2.2.20), (2.2.26) and the condition III we conclude that

$$\left|\Lambda(\lambda_\gamma) - \lambda_\gamma\right| < \left|c^\gamma\right| + \left|\sum_{\gamma_1} V_{\gamma_1-\gamma}\right| < \left(a_2 + ia_1\right)\left|(\gamma+t)_z\right| + M < K\cdot\left|\gamma+t\right|^{\alpha^1}.$$ (2.2.27)

Iterating the formula (2.2.26) (rather choosing $\left(\Phi_N, \phi_{\gamma_1}\right)$ in (2.2.27) $\gamma_1 = \gamma$ and applying

(2.2.26) to $\left(\Phi_N, \phi_{\gamma_1}\right)$ for $\gamma_1 \neq \gamma$) and considering $V_0 = 0$ (see (2.2.13)) we obtain that

$$\sum_{\gamma_1} V_{\gamma_1-\gamma}\left(\Phi_N, \phi_{\gamma_1}\right) = V_0\left(\Phi_N, \phi_\gamma\right) + \sum_{\gamma_1 \neq \gamma} V_{\gamma_1-\gamma}\left(\Phi_N, \phi_{\gamma_1}\right) = \sum_{\gamma_1} V_{\gamma_1-\gamma}\frac{\left(\Phi_N, N(h)\phi_{\gamma_1}\right)}{\Lambda_N - \lambda_{\gamma_1}} =$$

$$+\left[\sum_{\gamma_1 \neq \gamma} \frac{V_{\gamma_1-\gamma}V_{\gamma-\gamma_1}}{\Lambda_N - \lambda_{\gamma_1}} + \sum_{\gamma_1,\gamma_2 \neq \gamma} \frac{V_{\gamma_1-\gamma}V_{\gamma_2-\gamma_1}V_{\gamma-\gamma_2}}{\left(\Lambda_N - \lambda_{\gamma_1}\right)\left(\Lambda_N - \lambda_{\gamma_2}\right)} + ... + \sum_{\gamma_1 \neq \gamma} \frac{V_{\gamma_1-\gamma}V_{\gamma_2-\gamma_1}\cdots V_{\gamma-\gamma_{k-1}}}{\left(\Lambda_N - \lambda_{\gamma_1}\right)\cdots\left(\Lambda_N - \lambda_{\gamma_{k-1}}\right)}\right]\left(\Phi_N, \phi_\gamma\right) + ... +$$

$$+\sum_{\gamma_1 \neq \gamma} \frac{V_{\gamma_1-\gamma}V_{\gamma_2-\gamma_1}\cdots V_{\gamma_k-\gamma_{k-1}}}{\left(\Lambda_N - \lambda_{\gamma_1}\right)\cdots\left(\Lambda_N - \lambda_{\gamma_{k-1}}\right)}\left(\Phi_N, \phi_{\gamma_k}\right) + \sum_{\gamma_1 \neq \gamma} \frac{V_{\gamma_1-\gamma}c^{\gamma_1}}{\Lambda_N - \lambda_{\gamma_1}}\left(\Phi_N, \phi^{\gamma_1}\right) + \sum_{\gamma_1,\gamma_2 \neq \gamma} \frac{V_{\gamma_1-\gamma}V_{\gamma_2-\gamma_1}c^{\gamma_2}}{\left(\Lambda_N - \lambda_{\gamma_1}\right)\left(\Lambda_N - \lambda_{\gamma_2}\right)}\left(\Phi_N, \phi^{\gamma_2}\right) +$$

$$+...+\sum_{\gamma_1 \neq \gamma} \frac{V_{\gamma_1-\gamma}V_{\gamma_2-\gamma_1}\cdots V_{\gamma_{k-1}-\gamma_{k-2}}c^{\gamma_{k-1}}}{\left(\Lambda_N - \lambda_{\gamma_1}\right)\left(\Lambda_N - \lambda_{\gamma_2}\right)\cdots\left(\Lambda_N - \lambda_{\gamma_{k-1}}\right)}\left(\Phi_N, \phi^{\gamma_{k-1}}\right).$$

Now using the formulas (2.2.24) and (2.2.25) we have

$$c^\gamma\left(\Phi_N, \phi^\gamma\right) = c^\gamma \frac{\left(\Phi_N, N(h)\phi^\gamma\right)}{\Lambda_N - \lambda^\gamma} = \frac{c^\gamma b^\gamma}{\Lambda_N - \lambda^\gamma}\left(\Phi_N, \phi_\gamma\right) + \sum_{\gamma_1} \frac{V_{\gamma_1-\gamma}c^\gamma}{\Lambda_N - \lambda^\lambda}\left(\Phi_N, \phi^{\gamma_1}\right) = \left\{\frac{c^\gamma b^\gamma}{\Lambda - \lambda^\gamma} + \frac{c^\gamma b^\gamma}{\left(\Lambda - \lambda^\gamma\right)^2}\right.$$

$$\cdot\left[\sum_{\gamma_1}\frac{V_{\gamma_1-\gamma}V_{\gamma-\gamma_1}}{\left(\Lambda_N-\lambda^{\gamma_1}\right)}+\sum_{\gamma_1,\gamma_2}\frac{V_{\gamma_1-\gamma}V_{\gamma_2-\gamma_1}V_{\gamma-\gamma_2}}{\left(\Lambda_N-\lambda^{\gamma_1}\right)\left(\Lambda_N-\lambda^{\gamma_2}\right)}+\cdots+\sum_{\gamma_i}\frac{V_{\gamma_1-\gamma}V_{\gamma_2-\gamma_1}\cdots V_{\gamma_{k-1}-\gamma_{k-2}}V_{\gamma-\gamma_{k-1}}}{\left(\Lambda_N-\lambda^{\gamma_1}\right)\left(\Lambda_N-\lambda^{\gamma_2}\right)\cdots\left(\Lambda_N-\lambda^{\gamma_{k-1}}\right)}\right]\left(\Phi_N,\phi_\gamma\right)+$$

$$+\sum_{\gamma_i}\frac{c^\gamma V_{\gamma_1-\gamma}V_{\gamma_2-\gamma_1}\cdots V_{\gamma_{k-1}-\gamma_{k-2}}V_{\gamma_k-\gamma_{k-1}}}{\left(\Lambda_N-\lambda^{\gamma_1}\right)\left(\Lambda_N-\lambda^{\gamma_2}\right)\cdots\left(\Lambda_N-\lambda^{\gamma_{k-1}}\right)}\left(\Phi_N,\phi^{\gamma_k}\right)+\frac{c^\gamma}{\Lambda-\lambda^\gamma}\left[\sum_{\gamma_1\neq\gamma}\frac{V_{\gamma_1-\gamma}b^{\gamma_1}}{\Lambda_N-\lambda^{\gamma_1}}\left(\Phi_N,\phi_{\gamma_1}\right)+\right.$$

$$+\sum_{\gamma_1,\gamma_2\neq\gamma}\frac{V_{\gamma_1-\gamma}V_{\gamma_2-\gamma_1}b^{\gamma_2}}{\left(\Lambda_N-\lambda^{\gamma_1}\right)\left(\Lambda_N-\lambda^{\gamma_2}\right)}\left(\Phi_N,\phi_{\gamma_2}\right)+\cdots+\sum_{\gamma_i\neq\gamma}\frac{V_{\gamma_1-\gamma}V_{\gamma_2-\gamma_1}\cdots V_{\gamma_{k-1}-\gamma_{k-2}}b^{\gamma_{k-1}}}{\left(\Lambda_N-\lambda^{\gamma_1}\right)\left(\Lambda_N-\lambda^{\gamma_2}\right)\cdots\left(\Lambda_N-\lambda^{\gamma_{k-1}}\right)}\left(\Phi_N,\phi_{\gamma_{k-1}}\right)\right].$$

Denote:

$$S_k\left(\Lambda_N,\lambda^\gamma\right)=\sum_{\gamma_i\neq\gamma}\frac{V_{\gamma_1-\gamma}V_{\gamma_2-\gamma_1}\cdots V_{\gamma_{k-1}-\gamma_{k-2}}V_{\gamma-\gamma_{k-1}}}{\left(\Lambda_N-\lambda^{\gamma_1}\right)\left(\Lambda_N-\lambda^{\gamma_2}\right)\cdots\left(\Lambda_N-\lambda^{\gamma_{k-1}}\right)},$$

$$C_k\left(\Lambda_N,\lambda^\gamma\right)=\sum_{\gamma_i\neq\gamma}\frac{V_{\gamma_1-\gamma}V_{\gamma_2-\gamma_1}\cdots V_{\gamma_{k-1}-\gamma_{k-2}}V_{\gamma_k-\gamma_{k-1}}}{\left(\Lambda_N-\lambda^{\gamma_1}\right)\left(\Lambda_N-\lambda^{\gamma_2}\right)\cdots\left(\Lambda_N-\lambda^{\gamma_{k-1}}\right)}\left(\Phi_N,\phi^{\gamma_k}\right),$$

$$S_k\left(\Lambda_N,\lambda^\gamma\right)=\sum_{\gamma_i\neq\gamma}\frac{V_{\gamma_1-\gamma}V_{\gamma_2-\gamma_1}\cdots V_{\gamma_{k-1}-\gamma_{k-2}}V_{\gamma-\gamma_{k-1}}}{\left(\Lambda_N-\lambda_{\gamma_1}\right)\left(\Lambda_N-\lambda_{\gamma_2}\right)\cdots\left(\Lambda_N-\lambda_{\gamma_{k-1}}\right)},$$

$$C_k\left(\Lambda_N,\lambda^\gamma\right)=\sum_{\gamma_i\neq\gamma}\frac{V_{\gamma_1-\gamma}V_{\gamma_2-\gamma_1}\cdots V_{\gamma_{k-1}-\gamma_{k-2}}V_{\gamma_k-\gamma_{k-1}}}{\left(\Lambda_N-\lambda_{\gamma_1}\right)\left(\Lambda_N-\lambda_{\gamma_2}\right)\cdots\left(\Lambda_N-\lambda_{\gamma_{k-1}}\right)}\left(\Phi_N,\phi_{\gamma_k}\right).$$

Then

$$\sum_{\gamma_1}V_{\gamma_1-\gamma}\left(\Phi_N,\phi_{\gamma_1}\right)=\left(\sum_{n=1}^{k-1}S_n\left(\Lambda_N,\lambda^\gamma\right)\right)\left(\Phi_N,\phi_\gamma\right)+C_k\left(\Lambda_N,\lambda^\gamma\right)+$$

$$+\sum_{\gamma_1\neq\gamma}\frac{V_{\gamma_1-\gamma}c^{\gamma_1}}{\Lambda_N-\lambda_{\gamma_1}}\left(\Phi_N,\phi^{\gamma_1}\right)+\sum_{\gamma_1,\gamma_2\neq\gamma}\frac{V_{\gamma_1-\gamma}V_{\gamma_2-\gamma_1}c^{\gamma_2}}{\left(\Lambda_N-\lambda_{\gamma_1}\right)\left(\Lambda_N-\lambda_{\gamma_2}\right)}\left(\Phi_N,\phi^{\gamma_2}\right)+\cdots+$$

$$+\sum_{\gamma_i\neq\gamma;i=1,k-1}\frac{V_{\gamma_1-\gamma}V_{\gamma_2-\gamma_1}\cdots V_{\gamma_{k-1}-\gamma_{k-2}}c^{\gamma_{k-1}}}{\left(\Lambda_N-\lambda_{\gamma_1}\right)\left(\Lambda_N-\lambda_{\gamma_2}\right)\cdots\left(\Lambda_N-\lambda_{\gamma_{k-1}}\right)}\left(\Phi_N,\phi^{\gamma_{k-1}}\right);\tag{2.2.27a}$$

$$c^\gamma\left(\Phi_N,\phi^\gamma\right)=\left\{\frac{c^\gamma b^\gamma}{\Lambda-\lambda^\gamma}+\frac{c^\gamma b^\gamma}{\left(\Lambda-\lambda^\gamma\right)^2}\left(\sum_{n=1}^{k-1}S_n\left(\Lambda_N,\lambda^\gamma\right)\right)\right\}\left(\Phi_N,\phi_\gamma\right)+\frac{c^\gamma}{\Lambda-\lambda^\gamma}\left\{C_k\left(\Lambda_N,\lambda^\gamma\right)+\right.$$

$$+\sum_{\gamma_1\neq\gamma}\frac{V_{\gamma_1-\gamma}b^{\gamma_1}}{\Lambda_N-\lambda^{\gamma_1}}\left(\Phi_N,\phi_{\gamma_1}\right)+\sum_{\substack{\gamma_1;\\\gamma_2\neq\gamma}}\frac{V_{\gamma_1-\gamma}V_{\gamma_2-\gamma_1}b^{\gamma_2}}{\left(\Lambda_N-\lambda^{\gamma_1}\right)\left(\Lambda_N-\lambda^{\gamma_2}\right)}\left(\Phi_N,\phi_{\gamma_2}\right)+\cdots+$$

$$+\sum_{\substack{\gamma_i;i=1,k-2;\\\gamma_{k-1}\neq\gamma}}\frac{V_{\gamma_1-\gamma}V_{\gamma_2-\gamma_1}\cdots V_{\gamma_{k-1}-\gamma_{k-2}}b^{\gamma_{k-1}}}{\left(\Lambda_N-\lambda^{\gamma_1}\right)\left(\Lambda_N-\lambda^{\gamma_2}\right)\cdots\left(\Lambda_N-\lambda^{\gamma_{k-1}}\right)}\left(\Phi_N,\phi_{\gamma_{k-1}}\right)\right\}.\tag{2.2.276}$$

Let us simplify the expression

$$\sum_{\gamma_1 \neq \gamma} \frac{c^{\gamma_1} V_{\gamma_1-\gamma}}{\Lambda_N - \lambda^{\gamma_1}}\left(\Phi_N, \phi^{\gamma_1}\right) = \sum_{\gamma_2; \gamma_1 \neq \gamma} \frac{c^{\gamma_1} V_{\gamma_1-\gamma} V_{\gamma_2-\gamma_1}}{\left(\Lambda_N - \lambda_{\gamma_1}\right)\left(\Lambda_N - \lambda^{\gamma_1}\right)}\left(\Phi_N, \phi^{\lambda_2}\right) + \sum_{\gamma_1 \neq \gamma} \frac{c^{\gamma_1} V_{\gamma_1-\gamma} b^{\gamma_1}}{\left(\Lambda_N - \lambda_{\gamma_1}\right)\left(\Lambda_N - \lambda^{\gamma_1}\right)}\left(\Phi_N, \phi_{\gamma_1}\right) =$$

$$= \left\{ \left[\sum_{\gamma_1 \neq \gamma} \frac{c^{\gamma_1} b^{\gamma} V_{\gamma_1-\gamma} V_{\gamma-\gamma_1}}{\left(\Lambda_N - \lambda_{\gamma_1}\right)\left(\Lambda_N - \lambda^{\gamma_1}\right)\left(\Lambda_N - \lambda^{\gamma}\right)} + \cdots + \sum_{\substack{\gamma_i \neq \gamma_i; \\ i=1,k-2}} \frac{c^{\gamma_1} b^{\gamma} V_{\gamma_1-\gamma} V_{\gamma_2-\gamma_1} \cdots V_{\gamma-\gamma_{k-2}}}{\left(\Lambda_N - \lambda_{\gamma_1}\right)\left(\Lambda_N - \lambda^{\gamma_1}\right)\left(\Lambda_N - \lambda^{\gamma_1}\right)\cdots\left(\Lambda_N - \lambda^{\gamma_{k-2}}\right)} \right] + \right.$$

$$+ \left[\sum_{\gamma_1 \neq \gamma} \frac{c^{\gamma_1} b^{\gamma_1} V_{\gamma_1-\gamma} V_{\gamma-\gamma_1}}{\left(\Lambda_N - \lambda_{\gamma_1}\right)^2\left(\Lambda_N - \lambda^{\gamma}\right)} + \cdots + \sum_{\substack{\gamma_i \neq \gamma_i; \\ i=1,k-2}} \frac{c^{\gamma_1} b^{\gamma_1} V_{\gamma_1-\gamma} V_{\gamma_2-\gamma_1} \cdots V_{\gamma-\gamma_{k-2}}}{\left(\Lambda_N - \lambda_{\gamma_1}\right)^2\left(\Lambda_N - \lambda^{\gamma_1}\right)\left(\Lambda_N - \lambda_{\gamma_2}\right)\cdots\left(\Lambda_N - \lambda_{\gamma_{k-2}}\right)} \right] +$$

$$+ \left[\sum_{\gamma_1 \neq \gamma} \frac{c^{\gamma_1} b^{\gamma_2} V_{\gamma_1-\gamma} V_{\gamma_2-\gamma_1} V_{\gamma-\gamma_2}}{\left(\Lambda_N - \lambda^{\gamma_2}\right)\left(\Lambda_N - \lambda^{\gamma_1}\right)\left(\Lambda_N - \lambda_{\gamma_1}\right)\left(\Lambda_N - \lambda_{\gamma_2}\right)} + \cdots + \right.$$

$$\left. + \sum_{\substack{\gamma_i \neq \gamma_i; \\ i=1,k-2}} \frac{c^{\gamma_1} b^{\gamma_2} V_{\gamma_1-\gamma} V_{\gamma_2-\gamma_1} \cdots V_{\gamma-\gamma_{k-2}}}{\left(\Lambda_N - \lambda^{\gamma_2}\right)\left(\Lambda_N - \lambda^{\gamma_1}\right)\left(\Lambda_N - \lambda_{\gamma_1}\right)\left(\Lambda_N - \lambda_{\gamma_2}\right)\cdots\left(\Lambda_N - \lambda_{\gamma_{k-2}}\right)} \right] + \cdots +$$

$$\left. + \sum_{\substack{\gamma_i \neq \gamma_i; \\ i=1,k-2}} \frac{c^{\gamma_1} b^{\gamma_p} V_{\gamma_1-\gamma} V_{\gamma_2-\gamma_1} \cdots V_{\gamma-\gamma_{k-2}}}{(\Lambda_N - \lambda_{\gamma_1})(\Lambda_N - \lambda^{\gamma_p})(\Lambda_N - \lambda^{\gamma_1})(\Lambda_N - \lambda_{\gamma_2})\cdots(\Lambda_N - \lambda_{\gamma_{k-2}})} \right\}\left(\Phi_N, \phi_\gamma\right) +$$

$$+ \sum_{\substack{\gamma_1 \neq \gamma_i; \\ \gamma_i; i=2,k-2}} \frac{c^{\gamma_1} V_{\gamma_1-\gamma} V_{\gamma_2-\gamma_1} \cdots V_{\gamma_{k-1}-\gamma_{k-2}} V_{\gamma_k-\gamma_{k-1}}}{\left(\Lambda_N - \lambda^{\gamma_1}\right)\left(\Lambda_N - \lambda_{\gamma_1}\right)\cdots\left(\Lambda_N - \lambda^{\gamma_{k-1}}\right)}\left(\Phi_N, \phi^{\gamma_k}\right) +$$

$$+ \sum_{\substack{\gamma_i \neq \gamma_i; \\ i=1,k-1}} \frac{c^{\gamma_1} b^{\gamma_1} V_{\gamma_1-\gamma} V_{\gamma_2-\gamma_1} \cdots V_{\gamma_{k-1}-\gamma_{k-2}}}{\left(\Lambda_N - \lambda^{\gamma_1}\right)^2\left(\Lambda_N - \lambda_{\gamma_1}\right)\left(\Lambda_N - \lambda_{\gamma_2}\right)\cdots\left(\Lambda_N - \lambda_{\gamma_{k-2}}\right)}\left(\Phi_N, \phi_{\gamma_{k-1}}\right) + \cdots +$$

$$+ \sum_{\substack{\gamma_i \neq \gamma_i; \\ i=1,k-1}} \frac{c^{\gamma_1} b^{\gamma_p} V_{\gamma_1-\gamma} V_{\gamma_2-\gamma_1} \cdots V_{\gamma_{k-1}-\gamma_{k-2}}}{\left(\Lambda_N - \lambda^{\gamma_p}\right)\left(\Lambda_N - \lambda_{\gamma_1}\right)\left(\Lambda_N - \lambda^{\gamma_1}\right)\cdots\left(\Lambda_N - \lambda^{\gamma_{k-2}}\right)}\left(\Phi_N, \phi_{\gamma_{k-1}}\right) +$$

$$+ \cdots + \sum_{\gamma_1 \neq \gamma} \frac{V_{\gamma_1-\gamma} c^{\gamma_1} b^{\gamma_1} \cdots c^{\gamma_1} b^{\gamma_1}}{\left(\Lambda_N - \lambda_{\gamma_1}\right)^{\frac{p}{2}}\left(\Lambda_N - \lambda^{\gamma_1}\right)^{\frac{p}{2}}}\left(\Phi_N, \phi_{\gamma_1}\right). \tag{2.2.28}$$

From (2.2.26), (2.2.20), (2.2.21) and condition III we get

$$\left|\Lambda_N(\lambda_\gamma) - \lambda^{\gamma_1}\right| \geq \left|\lambda_\gamma - \lambda^{\gamma_1}\right| - \left|\Lambda_N(\lambda_\gamma) - \lambda_\gamma\right| > |\gamma + t|^\alpha - K|\gamma + t|^{\alpha^1} - 2\left[\gamma_1 + t, a\right]_z >$$

$$> |\gamma + t|^\alpha - 2M|\gamma + t|^{\alpha^1} > \frac{1}{2}|\gamma + t|^\alpha; \tag{2.2.29}$$

$$\left|\Lambda_N(\lambda_\gamma) - \lambda_\gamma\right| \geq \left|\lambda_\gamma - \lambda_{\gamma_1}\right| - \left|\Lambda_N(\lambda_\gamma) - \lambda_\gamma\right| > |\gamma + t|^\alpha - K|\gamma + t|^{\alpha^1} > \frac{1}{2}|\gamma + t|^\alpha. \tag{2.2.30}$$

Then

$$\sum_{\gamma_1 \neq \gamma} \frac{V_{\gamma_1 - \gamma} c^{\gamma_1} b^{\gamma_1} \cdots c^{\gamma_1} b^{\gamma_1}}{\left(\Lambda_N - \lambda_{\gamma_1}\right)^{\frac{p}{2}} \left(\Lambda_N - \lambda^{\gamma_1}\right)^{\frac{p}{2}}} = O\left(\left|\gamma + t\right|^{-(\alpha - \alpha^1)p}\right).$$

Therefore (2.2.28) may be written if the form

$$\sum_{\gamma_1 \neq \gamma} \frac{c^{\gamma_1} V_{\gamma_1 - \gamma}}{\Lambda_N - \lambda^{\gamma_1}} \left(\Phi_N, \phi^{\gamma_1}\right) = \left\{ \sum_{n=1}^{p} \frac{c^{\gamma_1} b^{\gamma}}{\left(\Lambda_N - \lambda_{\gamma_1}\right)\left(\Lambda_N - \lambda^{\gamma}\right)} \otimes S_n\left(\Lambda_N, \lambda^{\gamma}\right) + \sum_{n=1}^{p} \frac{c^{\gamma_1} b^{\gamma_1}}{\left(\Lambda_N - \lambda_{\gamma_1}\right)\left(\Lambda_N - \lambda^{\gamma_1}\right)} \otimes S_n\left(\Lambda_N, \lambda_{\gamma}\right) + \right.$$

$$+ \sum_{n=2}^{p} \frac{c^{\gamma_1} b^{\gamma_2}}{\left(\Lambda_N - \lambda_{\gamma_1}\right)\left(\Lambda_N - \lambda^{\gamma_2}\right)} \otimes S_n^{-1}\left(\Lambda_N, \lambda_{\gamma}\right) + \cdots + \sum_{n=p-1}^{p} \frac{c^{\gamma_1} b^{\gamma_{p-1}}}{\left(\Lambda_N - \lambda_{\gamma_1}\right)\left(\Lambda_N - \lambda^{\gamma_{p-1}}\right)} \otimes S_n^{p-2}\left(\Lambda_N, \lambda_{\gamma}\right) +$$

$$+ \frac{c^{\gamma_1} b^{\gamma_p}}{\left(\Lambda_N - \lambda_{\gamma_1}\right)\left(\Lambda_N - \lambda^{\gamma_p}\right)} \otimes S_n^{p-1}\left(\Lambda_N, \lambda_{\gamma}\right) \right\} \left(\Phi_N, \phi_{\gamma}\right) + O\left(\left|\gamma + t\right|^{-(\alpha - \alpha^1)p}\right). \qquad (2.2.31)$$

Here we used the denotations

$$\frac{c^{\gamma_1} b^{\gamma_k}}{\left(\Lambda_N - \lambda_{\gamma_1}\right)\left(\Lambda_N - \lambda^{\gamma_k}\right)} \otimes S_n^{k}\left(\Lambda_N, \lambda_{\gamma}\right) = \sum_{\substack{\gamma_i \neq \gamma_j \\ i=1, p-1}} \frac{c^{\gamma_1} b^{\gamma_k} V_{\gamma_1 - \gamma_2} V_{\gamma_2 - \gamma_3} \cdots V_{\gamma_p - \gamma_{p-1}} V_{\gamma - \gamma_p}}{\left(\Lambda_N - \lambda_{\gamma_1}\right)\left(\Lambda_N - \lambda^{\gamma_k}\right)\left(\Lambda_N - \lambda^{\gamma_1}\right) \cdots \left(\Lambda_N - \lambda^{\gamma_{k-1}}\right)\left(\Lambda_N - \lambda_{\gamma_k}\right) \cdots \left(\Lambda_N - \lambda_{\gamma_p}\right)}.$$

Similarly iterating the formula $\displaystyle\sum_{\gamma_1, \gamma_2 \neq \gamma} \frac{V_{\gamma_1 - \gamma} V_{\gamma_2 - \gamma_1} c^{\gamma_2}}{\left(\Lambda_N - \lambda_{\gamma_1}\right)\left(\Lambda_N - \lambda_{\gamma_2}\right)} \left(\Phi_N, \phi^{\gamma_2}\right)$ p-1 times we get:

$$\sum_{\gamma_1, \gamma_2 \neq \gamma} \frac{V_{\gamma_1 - \gamma} V_{\gamma_2 - \gamma_1} c^{\gamma_2}}{\left(\Lambda_N - \lambda_{\gamma_1}\right)\left(\Lambda_N - \lambda_{\gamma_2}\right)} \left(\Phi_N, \phi^{\gamma_2}\right) = \left\{ \sum_{n=2}^{p} \frac{c^{\gamma_2} b^{\gamma}}{\left(\Lambda_N - \lambda_{\gamma_2}\right)\left(\Lambda_N - \lambda^{\gamma}\right)} \otimes S_n^{1}\left(\Lambda_N, \lambda^{\gamma}\right) + \right.$$

$$+ \sum_{m=2}^{p} \sum_{n=2}^{p} \frac{c^{\gamma_2} b^{\gamma_m}}{\left(\Lambda_N - \lambda_{\gamma_2}\right)\left(\Lambda_N - \lambda^{\gamma_m}\right)} \otimes S_n^{m-1}\left(\Lambda_N, \lambda_{\gamma}\right) \right\} \left(\Phi_N, \phi_{\gamma}\right) + O\left(\left|\gamma + t\right|^{-(\alpha - \alpha^1)p}\right). \qquad (2.2.32)$$

From this

$$\sum_{\substack{\gamma_i \neq \gamma_j \\ i=1, k-1}} \frac{c^{\gamma} V_{\gamma_1 - \gamma} V_{\gamma_2 - \gamma_1} \cdots V_{\gamma_{k-1} - \gamma_{k-2}}}{\left(\Lambda_N - \lambda_{\gamma_1}\right)\left(\Lambda_N - \lambda_{\gamma_2}\right) \cdots \left(\Lambda_N - \lambda_{\gamma_{k-2}}\right)} \left(\Phi_N, \phi^{\gamma_{k-1}}\right) =$$

$$= \left\{ \sum_{n=k-1}^{p} \frac{c^{\gamma_{k-1}} b^{\gamma}}{\left(\Lambda_N - \lambda_{\gamma_{k-1}}\right)\left(\Lambda_N - \lambda^{\gamma}\right)} \otimes S_n^{k-2}\left(\Lambda_N, \lambda^{\gamma}\right) + \right.$$

$$+ \sum_{m=k-1}^{p} \sum_{n=k-1}^{p} \frac{c^{\gamma_{k-1}} b^{\gamma_m}}{\left(\Lambda_N - \lambda_{\gamma_{k-1}}\right)\left(\Lambda_N - \lambda^{\gamma_m}\right)} \otimes S_n^{m-1}\left(\Lambda_N, \lambda_{\gamma}\right) \right\} \left(\Phi_N, \phi_{\gamma}\right) + O\left(\left|\gamma + t\right|^{-(\alpha - \alpha^1)p}\right). \qquad (2.2.33)$$

Substituting (2.2.31), (2.2.32) and (2.2.33) into (2.2.26) we obtain

$$\left(\Lambda_N - \lambda_{\gamma}\right)\left(\Phi_N, \phi_{\gamma}\right) = \left\{ \sum_{n=1}^{k-1} S_n\left(\Lambda_N, \lambda_{\gamma}\right) + \frac{c^{\gamma} b^{\gamma}}{\left(\Lambda_N - \lambda^{\gamma}\right)^{2}} \left(\sum_{n=1}^{k-1} S_n\left(\Lambda_N, \lambda_{\gamma}\right)\right) + \right.$$

$$+\sum_{r=2}^{k-1}\sum_{n=r}^{p}\frac{c^{\gamma_r}b^{\gamma}}{\left(\Lambda_N-\lambda_{\gamma_r}\right)\left(\Lambda_N-\lambda^{\gamma}\right)}\otimes S_n^{r-1}\left(\Lambda_N,\lambda^{\gamma}\right)+$$

$$++\sum_{r=1}^{k-1}\sum_{m=r}^{p}\sum_{n=r}^{p}\frac{c^{\gamma_r}b^{\gamma_m}}{\left(\Lambda_N-\lambda_{\gamma_r}\right)\left(\Lambda_N-\lambda^{\gamma_m}\right)}\otimes S_n^{m-1}\left(\Lambda_N,\lambda_\gamma\right)+$$

$$+\frac{c^{\gamma}}{\Lambda_N-\lambda_\gamma}\left[\sum_{r=1}^{k-1}\sum_{n=r}^{p}\frac{b^{\gamma_r}}{\Lambda_N-\lambda^{\gamma_r}}\otimes S_n^{r-1}\left(\Lambda_N,\lambda_\gamma\right)+\right.$$

$$+\sum_{r=1}^{k-1}\sum_{n=r}^{p}\frac{c^{\gamma_r}b^{\gamma_r}b^{\gamma}}{\left(\Lambda_N-\lambda_{\gamma_r}\right)\left(\Lambda_N-\lambda^{\gamma_r}\right)\left(\Lambda_N-\lambda^{\gamma}\right)}\otimes S_n^{r-1}\left(\Lambda_N,\lambda^{\gamma}\right)+$$

$$\left.+\sum_{r=1}^{k-1}\sum_{n=r}^{p}\sum_{m=r}^{p}\frac{b^{\gamma_r}c^{\gamma_r}b^{\gamma_m}}{\left(\Lambda_N-\lambda^{\gamma_r}\right)\left(\Lambda_N-\lambda_{\gamma_r}\right)\left(\Lambda_N-\lambda^{\gamma_m}\right)}\otimes S_n^{m-1}\left(\Lambda_N,\lambda^{\gamma}\right)\right\}\left(\Phi_N,\phi_\gamma\right)+O\left(|\gamma+t|^{-(\alpha-\alpha^1)p}\right); \quad (2.2.37)$$

From the last we get:

$$\left(\Lambda_N-\lambda_\gamma\right)\left(\Phi_N,\phi_\gamma\right)=\left(\sum_{n=1}^{p-1}F_n\right)\left(\Phi_N,\phi_\gamma\right)+O\left(|\gamma+t|^{-(\alpha-\alpha^1)p}\right), \quad (2.2.38)$$

where F_n consists of the following terms

$$E_{i+j}(\Lambda_N)=\frac{c^{\gamma_{k_i}}\cdots c^{\gamma_{k_i}}b^{\gamma_{m_i}}\cdots b^{\gamma_{n_i}}V_{n-\gamma_1}V_{\gamma_2-\gamma_1}\cdots V_{r-\gamma_{m+j}}}{\left(\Lambda_N-\lambda_{\gamma_{k_i}}\right)\cdots\left(\Lambda_N-\lambda_{\gamma_{k_i}}\right)\left(\Lambda_N-\lambda^{\gamma_{n_i}}\right)\cdots\left(\Lambda_N-\lambda^{\gamma_{n_i}}\right)\left(\Lambda_N-\lambda_{\gamma_i}\right)\cdots\left(\Lambda_N-\lambda_{\gamma_{m+j}}\right)}. \quad (2.2.39)$$

Now we estimate these terms. Since $c^{\gamma_{k_i}}<M|\gamma+t|^{\alpha^1}$, $b^{\gamma_{n_j}}<N|\gamma+t|^{\alpha^1}$ (see (2.2.20a)) and

$$\left|\Lambda_N(\lambda_\gamma)-\lambda^{\gamma_{n_j}}\right|>\frac{1}{2}|\gamma+t|^{\alpha}; \qquad \left|\Lambda_N(\lambda_\gamma)-\lambda_{\gamma_{k_i}}\right|>\frac{1}{2}|\gamma+t|^{\alpha} \text{ (see (2.2.29) и (2.2.30)), we have}$$

$$E_p\left(\Lambda_N\right)=O\left(|\gamma+t|^{-(\alpha-\alpha^1)p}\right). \quad (2.2.40)$$

Since $V(h)$ is a trigonometric polynomial, $\sum_{\gamma\in Q}V_\gamma<c$ and it is easy to prove that

$$F_n=O\left(|\gamma+t|^{-p(\alpha-\alpha^1)}\right) \quad (2.2.41)$$

Therefore dividing (2.2.38) by $\left(\Phi_N,\varphi_\gamma\right)$ we get

$$\Lambda_N\left(\lambda_\gamma\right)=\lambda_\gamma+\sum_{n=1}^{m-1}F_n\left(\Lambda_N\right)+O\left(|\gamma+t|^{-(\alpha-\alpha^1)m}\right). \quad (2.2.42)$$

Finally as in $[6]$ replacing Λ_N by $\lambda_\gamma+\sum_{n=1}^{m-1}F_n\left(\lambda_\gamma\right)$ in the dominator $E_p\left(\Lambda_N\right)$ of the formula

(2.2.42) we come to

$$\Lambda_N\left(\lambda_\gamma\right)=\lambda_\gamma+\sum_{n=1}^{m-1}\overline{F}_n\left(\lambda_\gamma,\lambda^\gamma,V\left(h\right)\right)+O\left(\left|\gamma+t\right|^{-\left(\alpha-\alpha^1\right)m}\right). \tag{2.2.43}$$

From all these considerations the it follows validity of the following

Theorem 2.2.3. If the conditions I – III are satisfied, $\gamma+t\in V_v\left(\alpha\right)$ and $\left|\gamma+t\right|$ is large enough, then the corresponding eigenvalue $\Lambda_N\left(\lambda_\gamma\right)$ of the operator $H_t\left(a,V\left(h\right)\right)$ satisfies the asymptotic formula (2.2.43), where $\overline{F}_n\left(\lambda_\gamma,\lambda^\gamma,V\left(h\right)\right)$ is explicitly expressed by λ_γ, λ^γ, $V\left(h\right)$ and has an order $O\left(\left|\gamma+t\right|^{-\left(\alpha-\alpha^1\right)n}\right)$.

Remark: As one can see from the proof of the theorem, if $\Lambda_N\left(\lambda_\gamma\right)$ is such that $\left|\left(\Phi_N,\phi_\gamma\right)\right|>\left|\gamma+t\right|^{-\left(\alpha-\alpha^1\right)s}$, then $\Lambda_N\left(\lambda_\gamma\right)$ satisfies the asymptotic formula obtained by replacement $O\left(\left|\gamma+t\right|^{-\left(\alpha-\alpha^1\right)n}\right)$ by $O\left(\left|\gamma+t\right|^{-\left(\alpha-\alpha^1\right)\left(n-s\right)}\right)$ in (2.2.43). The similar formula is true also for $\Lambda_N\left(\lambda^\gamma\right)$.

Now let us discuss the question: ***Does exist an eigenvalue Λ_N of the operator $H_t\left(a,V\left(h\right)\right)$ in each above neighborhood of the eigenvalue λ_γ?***

In this purpose we need the following lemma proved in $\left[6\right]$.

Lemma: Foe the arbitrary elements $h,h_1,h_2,...,h_k\in H$ there exists a number N such that

$$\left|\left(h,\Phi_N\right)\right|^2\geq\frac{1}{k}\sum_{n=1}^k\left|\left(h_n,\Phi_N\right)\right|^2>\frac{1}{k}\left|\left(h_n,\Phi_N\right)\right|^2,\quad\text{where }n=\overline{1,k},\ \left\|h_n\right\|=1,\ \left\|h\right\|=1;\text{ here }\Phi_1,\Phi_2,...\text{ is any}$$

orthonormal basis (for instance eigenfunctions of the self adjoint operator $H_t\left(a,V\left(h\right)\right)$) in H.

As vectors $h,h_1,h_2,...,h_k$ we take

$$\varphi_\gamma,\ \overline{C}_m\left(\lambda_\gamma\right),\ \frac{\partial}{\partial x}\overline{C}_m\left(x\right)\bigg|_{x=\lambda_\gamma}\equiv\overline{C}_m^1\left(\lambda_\gamma\right),...,\ \frac{\partial^{k-1}}{\partial x^{k-1}}\overline{C}_m\left(x\right)\bigg|_{x=\lambda_\gamma}\equiv\overline{C}_m^{k-1}\left(\lambda_\gamma\right),$$

where $\overline{C}_m\left(x\right)=\sum_{k,n_j}\dfrac{c^{\gamma_k}\cdots c^{\gamma_k}b^{\gamma_n}\cdots b^{\gamma_{n_j}}V_{\gamma_1-\gamma}V_{\gamma_2-\gamma_1}\cdots V_{\gamma-\gamma_{m-i-j}}}{\left(x-\lambda_{\gamma_{n_1}}\right)\cdots\left(x-\lambda_{\gamma_{n_k}}\right)\left(x-\lambda^{\gamma_{n_1}}\right)\cdots\left(x-\lambda^{\gamma_{n_j}}\right)\left(x-\lambda_{\gamma_1}\right)\cdots\left(x-\lambda_{\gamma_{m-i}}\right)}\phi_{\gamma_{m-i-j}}.$ \tag{2.2.45}

Consider the known fact

$$\sum_{N:\left|\Lambda_N-\lambda_\gamma\right|\geq2M}\left|\left(\Phi_N,\phi_\gamma\right)\right|^2=\sum_{N:\left|\Lambda_N-\lambda_\gamma\right|\geq2M}\frac{\left|\left(\Phi_N,N\left(h\right)\phi_\gamma\right)\right|^2}{\left|\Lambda_N-\lambda_\gamma\right|^2}\leq\frac{\left\|N\left(h\right)\phi_\gamma\right\|}{4M}<\frac{1}{4},\qquad\text{here: }M=\left\|N\left(h\right)\phi_\gamma\right\|.$$

It is obvious that the number of the eigenvalues Λ_N from the interval $\left[\lambda_\gamma - 2M, \lambda_\gamma + 2M\right]$ is less than $\lambda_\gamma^{k\left(a-a^1\right)}$ by $\lambda_\gamma \to \infty$. Denote by $A\left(\lambda_\gamma\right)$ the set of indexes N such that $\Lambda_N \in \left[\lambda_\gamma - 2M, \lambda_\gamma + 2M\right]$ and $\left|\left(\Phi_N, \phi_\gamma\right)\right| > \frac{1}{2}\lambda_\gamma^{-\frac{1}{2}k\left(a-a^1\right)}$.

Then we get

$$\sum_{N\in A\left(\lambda_\gamma\right)} \left|\left(\Phi_N, \phi_\gamma\right)\right|^2 = 1 - \sum_{\substack{N:\left|\Lambda_N - \lambda_\gamma\right| > 2M}} \left|\left(\Phi_N, \phi_\gamma\right)\right|^2 - \sum_{\substack{N:\left|\Lambda_N - \lambda_\gamma\right| < 2M \\ \left|\left(\Phi_N, \phi_\gamma\right)\right| < \frac{1}{2}\lambda_\gamma^{-\frac{1}{2}k\left(a-a^1\right)}}} \left|\left(\Phi_N, \phi_\gamma\right)\right|^2 \geq \frac{1}{2} . \tag{2.2.47}$$

From this and the proof of the lemma we obtain that there exists N from $A(\lambda_\gamma)$ for which

$$\left|\left(\Phi_N, \phi_\gamma\right)\right| \geq \frac{1}{4k}\left|\left(\overline{C}_m^j\left(\lambda_\gamma\right), \Phi_N\right)\right| \cdot \frac{1}{\left\|\overline{C}_m^j\left(\lambda_\gamma\right)\right\|} , \tag{2.2.48}$$

because otherwise it would follow that

$$\frac{1}{2} \leq \sum_{N\in A\left(\lambda_\gamma\right)} \left|\left(\Phi_N, \phi_\gamma\right)\right|^2 \leq \sum_{N\in A\left(\lambda_\gamma\right)} \left|\left(\Phi_N, \frac{1}{4k}\frac{\overline{C}_m^j\left(\lambda_\gamma\right)}{\left\|\overline{C}_m^j\left(\lambda_\gamma\right)\right\|}\right)\right|^2 < \frac{1}{4} .$$

Now similarly to $\left[6\right]$ may be proved that from (2.2.48) follows the existence of $\Lambda_N\left(\lambda_\gamma\right)$ that satisfies the formula (2.2.43).

Theorem 2.2.4. If the conditions of Theorem 2.2.3 are satisfied, then for any λ_γ there exists the corresponding eigenvalues $\Lambda_N\left(\lambda_\gamma\right)$ of the operator $H_t\left(a, V\left(h\right)\right)$ that satisfies the asymptotic formula (2.2.43).

2.3. Asymptotic formulas for the non-resonance
eigenvalues of the Pauli operator

In the chapter we consider the Pauli type operator $H_t^l\left(a, V\left(h\right)\right)$ generated in $L_2\left(F\right) \times L_2\left(F\right)$ by the expression

$$H^l\left(a, V\left(h\right)\right) = \left(\left(-i\nabla - a\right)^{2l} + V\left(h\right)\right) \cdot I + \sigma \cdot B, \quad l > 1 \tag{2.3.1}$$

and boundary conditions

$$u\left(h + \omega_j\right) = e^{2\pi i t_j} u\left(h\right), \qquad j = \overline{1,3} , \tag{2.3.2}$$

where $h=(x,y,z)\in R^3$, $B=[\nabla,a]$ is a magnet field generated by the vector potential $a=(a_1,a_2,a_3)$, $B=(B_1,B_2,B_3)$,

$$I=\begin{pmatrix}1 & 0\\ 0 & 1\end{pmatrix},\quad B_1=\begin{vmatrix}\dfrac{\partial}{\partial y} & \dfrac{\partial}{\partial z}\\ a_2 & a_3\end{vmatrix},\quad B_2=\begin{vmatrix}\dfrac{\partial}{\partial z} & \dfrac{\partial}{\partial x}\\ a_3 & a_1\end{vmatrix},\quad B_3=\begin{vmatrix}\dfrac{\partial}{\partial x} & \dfrac{\partial}{\partial y}\\ a_1 & a_2\end{vmatrix},\quad (2.3.3)$$

$$\sigma_1=\begin{pmatrix}0 & 1\\ 1 & 0\end{pmatrix},\quad \sigma_2=\begin{pmatrix}0 & -i\\ i & 0\end{pmatrix},\quad \sigma_3=\begin{pmatrix}1 & 0\\ 0 & -1\end{pmatrix},\quad (2.3.4)$$

F is a fundamental domain of some lattice $\Omega=\{m_1\omega_1+m_2\omega_2+m_3\omega_3:m_1,m_2,m_3\in Z\}$, i.e. parallelepiped, $t=t_1\gamma^1+t_2\gamma^2+t_3\gamma^3$, and $\gamma^1,\gamma^2,\gamma^3$ are biorthogonal to $\omega_1,\omega_2,\omega_3$ vectors, F^* is fundamental domain of the lattice $\Gamma=\{n_1\gamma^1+n_2\gamma^2+n_3\gamma^3:n_1,n_2,n_3\in Z\}$, $V(h)$ - periodic, smooth enough function.

Considering (3) and (4) the operator $H^l(a,V(h))$ may be written as follows

$$H^l(a,V(h))=\begin{pmatrix}(-i\nabla-a)^{2l}+V(h)+[\nabla,a]_z & [\nabla,a]_x-i[\nabla,a]_y\\ [\nabla,a]_x+i[\nabla,a]_y & (-i\nabla-a)^{2l}+V(h)-[\nabla,a]_z\end{pmatrix}.\quad (2.3.5)$$

Here we divide the operator $H^l(a,V(h))$ into two parts на две части, and consider the first part as an unperturbed operator and the second one as a perturbation. This scheme has a simple structure and it is applicable when.

Let us represent $H^l(a,V(h))$ in the form

$$H^l(a,V(h))=\begin{pmatrix}(-i\nabla-a)^{2l}+[\nabla,a]_z & 0\\ 0 & (-i\nabla-a)^{2l}-[\nabla,a]_z\end{pmatrix}+\begin{pmatrix}V(h) & [\nabla,a]_x-i[\nabla,a]_y\\ [\nabla,a]_x+i[\nabla,a]_y & V(h)\end{pmatrix}.$$

Denote by $M_t^l(a)$ the operator generated in $L_2(F)\times L_2(F)$ by the expression

$$M^l(a)=\begin{pmatrix}(-i\nabla-a)^{2l}+[\nabla,a]_z & 0\\ 0 & (-i\nabla-a)^{2l}-[\nabla,a]_z\end{pmatrix}\quad (2.3.7)$$

and boundary conditions (2).

By $N(h)$ we define the expession

$$N(h)=\begin{pmatrix}V(h) & [\nabla,a]_x-i[\nabla,a]_y\\ [\nabla,a]_x+i[\nabla,a]_y & V(h)\end{pmatrix}.\quad (2.3.8)$$

As in chapter 2 one can easily prove that eigenvalues and eigenfunctions of the operator $M_t^l(a)$ indeed are

$$\lambda_\gamma=(-i|\gamma+t|-a)^{2l}-i[\gamma+t,a]_z,\quad (2.3.9)$$

$$\lambda^{\gamma} = \left(-i|\gamma+t|-a\right)^{2l} + i[\gamma+t,a]_z ; \qquad (2.3.10)$$

$$\phi_{\gamma}(h) = \begin{pmatrix} 0 \\ e^{i(\gamma+t,h)} \end{pmatrix}, \quad \text{где} \quad \gamma \in \Gamma, \quad t \in F, \qquad (2.3.11)$$

$$\phi^{\gamma}(h) = \begin{pmatrix} e^{i(\gamma+t,h)} \\ 0 \end{pmatrix}, \quad \text{где} \quad \gamma \in \Gamma, \quad t \in F. \qquad (2.3.12)$$

Therefore the spectrum of the operator $M_t^l(a)$ consists of λ_{γ} and λ^{γ}, where $\gamma \in \Gamma$. Now we find asymptotic formulas for the eigenvalues of the operator $H^l(a, V(h))$. Again for the sake of simplicity we assume that $V(h)$ is a trigonometric polynomial

$$V(h) = \sum_{\gamma \in Q} V_{\gamma} e^{i(\gamma,h)}, \quad \text{where} \quad V_{\gamma} = \int_F V(h) e^{-i(\gamma,h)} dh, \quad Q = \{\gamma \in \Gamma : V_{\gamma} \neq 0\}.$$

Additionally suppose that

$$\int_F V(h) dh = 0. \qquad (2.3.13)$$

Denote

$$V_{\gamma}(\alpha) = \left\{ h \in R^3 : \left| |h|^{2l} - |h+\gamma|^{2l} \right| < |h|^{\alpha l} \right\}, \quad \alpha l > 1,$$

$$V^{(m)}(\alpha) = \bigcup_{\gamma \in Q^m} V_{\gamma}(\alpha), \quad \overline{V^{(m)}}(\alpha) = R^3 \setminus V^{(m)}(\alpha),$$

where

$$Q^{(m)} = \left\{ \gamma : \ \gamma = \gamma_1 + \gamma_2 + \dots + \gamma_p, \quad \gamma_i \in Q, \quad i = \overline{1,p} \quad p < m \right\}.$$

Now we define the eigenvalues associates with c λ_{γ} through $N(h)$. For this purpose we calculate $N(h)\phi_{\gamma}(h)$ and $N(h)\phi^{\gamma}(h)$:

$$N(h)\phi_{\gamma}(h) = \begin{pmatrix} \left([|\gamma+t|,a]_x - i[|\gamma+t|,a]_y\right) \cdot e^{i(\gamma+t,h)} \\ \sum_{\gamma_1 \in Q} V_{\gamma_1} e^{i(\gamma+\gamma_1+t,h)} \end{pmatrix},$$

$$N(h)\phi^{\gamma}(h) = \begin{pmatrix} \sum_{\gamma_1 \in Q} V_{\gamma_1} e^{i(\gamma+\gamma_1+t,h)} \\ \left([|\gamma+t|,a]_x + i[|\gamma+t|,a]_y\right) \cdot e^{i(\gamma+t,h)} \end{pmatrix}.$$

From this we get

$$\left(N(h) \cdot \phi_{\gamma}(h), \phi_{\beta}(h)\right) = \begin{cases} 0, & \beta \neq \gamma+\gamma_1 \\ \sum_{\gamma_1 \in Q} V_{\gamma_1}, & \beta = \gamma+\gamma_1 \end{cases}; \qquad (2.3.16)$$

$$\left(N(h)\cdot\phi^{\gamma}(h),\phi^{\beta}(h)\right)=\begin{cases}0, & \beta\neq\gamma+\gamma_1\\ \sum\limits_{\gamma_1\in Q}V_{\gamma_1}, & \beta=\gamma+\gamma_1\end{cases}. \tag{2.3.17}$$

These formulas show that λ_{γ} is associated with $\lambda_{\gamma+\gamma_1}$ and λ^{γ} - with $\lambda^{\gamma+\gamma_1}$ through $N(h)$.

Then it is clear that

$$\left(N(h)\cdot\phi_{\gamma}(h),\phi^{\beta}(h)\right)=\begin{cases}0, & \beta\neq\gamma\\ [\gamma+t,a]_x-i[\gamma+t,a]_y, & \beta=\gamma\end{cases}, \tag{2.3.18}$$

$$\left(N(h)\cdot\phi^{\gamma}(h),\phi^{\beta}(h)\right)=\begin{cases}0, & \beta\neq\gamma\\ [\gamma+t,a]_x+i[\gamma+t,a]_y, & \beta=\gamma\end{cases}. \tag{2.3.19}$$

If to denote

$$c^{\gamma}=[\gamma+t,a]_x-i[\gamma+t,a]_y, \quad b^{\gamma}=[\gamma+t,a]_x+i[\gamma+t,a]_y, \tag{2.3.20}$$

then one can show that

$$c^{\gamma}=O\big(|\gamma+t|\big), \qquad b^{\gamma}=O\big(|\gamma+t|\big). \tag{2.3.20a}$$

Suppose that $\gamma+t\in\overline{\overline{V^{(m)}}}(\alpha)$. Then for any $\gamma_1\in Q^{(m)}$ the following relations are true

$$\left|\lambda_{\gamma}-\lambda_{\gamma+\gamma_1}\right|=\left|\left(-i|\gamma+t|-a\right)^{2l}-i[\gamma+t,a]_z-\left(-i|\gamma+\gamma_1+t|-a\right)^{2l}+i[\gamma+\gamma_1+t,a]_z\right|\geq$$

$$\geq\left|\,|\gamma+t|^{2l}-|\gamma+\gamma_1+t|^{2l}\right|+c_1|a|\left|\,|\gamma+t|^{2l}-|\gamma+\gamma_1+t|^{2l}\right|+c_2|a|^2\left|\,|\gamma+t|^{2l-1}-|\gamma+\gamma_1+t|^{2l-1}\right|+\ldots+$$

$$+c_{2l-1}|a|^{2l-1}\left|\,|\gamma+t|-|\gamma+\gamma_1+t|\right|-\left|[\gamma_1,a]_z\right|\geq k|\gamma+t|^{\alpha l}-\left|[\gamma_1,a]_z\right|.$$

The condition $\gamma+t\in\overline{\overline{V^{(m)}}}(\alpha)$ means that $\gamma+t\notin V_{\gamma_1}(\alpha)$ i.e.

$$\left|\,|\gamma+t|^2-|\gamma+\gamma_1+t|^2\right|>|\gamma+t|^{\alpha l}, \text{ где } 0<\alpha<1.$$

By elementary calculations we arrive to

$$\left|[\gamma_1,a]_z\right|<C|\gamma_1|\cdot|a|<M\cdot|\gamma+t|<\frac{1}{2}|\gamma+t|^{\alpha l}.$$

From the last we get

$$\left|\lambda_{\gamma}-\lambda_{\gamma+\gamma_1}\right|>k|\gamma+t|^{\alpha l}. \tag{2.3.21}$$

Thus, if $\gamma+t\in\overline{\overline{V^{(m)}}}(\alpha)$, then λ_{γ} is a non-resonance eigenvalue of the operator $M_t^l(a)$.

By the similar way may be shown that

$$\left|\lambda^{\gamma}-\lambda^{\gamma+\gamma_1}\right|>k|\gamma+t|^{\alpha l}. \tag{2.3.21a}$$

The principle role in proof of asymptotic formulas plays the foolowing known formula

$$(\Lambda_N-\lambda_{\gamma})(\Phi_N,\varphi_{\gamma})=(\Phi_N,N(h)\varphi_{\gamma}) \tag{2.3.23}$$

Substituting into (2.3.23) the decomposition

$$N(h)\varphi_\gamma(h) = \sum_\beta (N(h)\varphi_\gamma, \varphi_\beta) p_\beta(h) + \sum_\beta (N(h)\varphi_\gamma, \varphi^\beta) p^\beta(h)$$

and considering the (2.3.16), (2.3.17), (2.3.18) и (2.3.19) we obtain

$$N(h)\phi_\gamma(h) = \sum_{\gamma_1} V_{\gamma_1 - \gamma} \phi_{\gamma_1}(h) + c^\gamma \phi^\gamma(h), \tag{2.3.24}$$

$$N(h)\phi^\gamma(h) = \sum_{\gamma_1} V_{\gamma_1 - \gamma} \phi^{\gamma_1}(h) + b^\gamma \phi_\gamma(h). \tag{2.3.25}$$

Then from (2.3.23) we have

$$(\Lambda - \lambda_\gamma)(\Phi, \phi_\gamma) = \sum_{\gamma_1} V_{\gamma_1 - \gamma}(\Phi(h), \phi_{\gamma_1}(h)) + c^\gamma (\Phi(h), \phi^\gamma(h)). \tag{2.3.26}$$

From the formulas (2.3.20a) and (2.3.26) may be obtained that

$$\left|\Lambda(\lambda_\gamma) - \lambda_\gamma\right| < |c^\gamma| + \left|\sum_{\gamma_1} V_{\gamma_1 - \gamma}\right| < K \cdot |\gamma + t|. \tag{2.3.27}$$

Iterating (2.3.26) (i.e. choosing (Φ, ϕ_{γ_1}) by $\gamma_1 = \gamma$ and applying (2.3.26) to (Φ, ϕ_{γ_1}) by $\gamma_1 \neq \gamma$) and considering $V_0 = 0$ (see (2.3.13)) we obtain

$$(\Lambda - \lambda_\gamma)(\Phi, \phi_\gamma) = \left(\sum_{n=1}^{p-1} F_n\right)(\Phi, \phi_\gamma) + O\left(|\gamma + t|^{-p(\alpha l - 1)}\right),$$

Here F_n consists of the following terms:

$$E_{l+j}(\Lambda) = \frac{c(\gamma_{k_1})...c(\gamma_{k_l}) b(\gamma_{n_1})...b(\gamma_{n_j}) V_{\gamma_1 - \gamma} V_{\gamma_2 - \gamma_1} \cdots V_{\gamma - \gamma_{m-l-j}}}{(\Lambda - \lambda_{\gamma_{k_1}}) \cdots (\Lambda - \lambda_{\gamma_{k_l}})(\Lambda - \lambda^{\gamma_{n_1}}) \cdots (\Lambda - \lambda^{\gamma_{n_j}})(\Lambda - \lambda_{\gamma_1}) \cdots (\Lambda - \lambda_{\gamma_{m-l-j}})},$$

Now let us estimate these terms. Since $\gamma + t \in \overline{V^{(m)}}(\alpha)$ и $\Lambda \equiv \Lambda(\lambda_\gamma)$ we can write

$$\left|\Lambda(\lambda_\gamma) - \lambda_{\gamma_1}\right| \geq \left|\lambda_\gamma - \lambda_{\gamma_1}\right| - \left|\Lambda_N(\lambda_\gamma) - \lambda_\gamma\right| > K |\gamma + t|^{\alpha l},$$

$$\left|\Lambda(\lambda_\gamma) - \lambda^{\gamma_1}\right| = \left|\Lambda(\lambda_\gamma) - \lambda_{\gamma_1}\right| - \left|\lambda_{\gamma_1} - \lambda^{\gamma_1}\right| > K |\gamma + t|^{\alpha l} - 2[\gamma + t, a]_z > M |\gamma + t|^{\alpha l}. \tag{2.3.30}$$

Consideration this and (2.3.20a) gives

$$E_p(\Lambda(\lambda_\gamma)) = O\left(|\gamma + t|^{-m\alpha l + p}\right).$$

since $V(h)$ is a trigonometric polynomial i.e.. $\left|\sum_{\gamma \in Q} V_\gamma\right| < c$. In [3] and [4] is proved that

$$F_n = O\left(|\gamma + t|^{-\alpha l m + n}\right). \tag{2.3.31}$$

Therefore dividing (38) by (Φ, ϕ_γ), we get

$$\Lambda(\lambda_{\gamma}) = \lambda_{\gamma} + \sum_{k=1}^{m-1} F_k\left(\Lambda\left(\lambda_{\gamma}\right)\right) + O\left(\left|\gamma + t\right|^{m(1-\alpha l)}\right). \tag{2.3.32}$$

Finally, as in $[2]$ in the formula (2.3.32) replacing Λ_N by $\lambda_{\gamma} + \sum_{n=1}^{m-1} F_n\left(\lambda_{\gamma}\right)$ in the denominate

$E_p\left(\Lambda_N\right)$ we get

$$\Lambda\left(\lambda_{\gamma}\right) = \lambda_{\gamma} + \sum_{k=1}^{m-1} \overline{F_k}\left(\lambda_{\gamma}, \lambda^{\gamma}, V(x)\right) + O\left(\left|\gamma + t\right|^{m(1-\alpha l)}\right). \tag{2.3.33}$$

From above considerations follows that the following theorem has been proved.

Theorem 2.3.1. If $\gamma + t \in \overline{\overline{V}}^m(\alpha)$, i.e. λ_{γ} is non-resonance eigenvalue of the operator $M_t^l(a)$, then the corresponding eigenvalue $\Lambda(\lambda_{\gamma})$ of the operator $H_t^l(a, V(h))$ (i.e. Pauli type operator) satisfies the asymptotic formula (2.3.43), where $\overline{F_n}\left(\lambda_{\gamma}, \lambda^{\gamma}, V(h)\right)$ is explicitly expressed through λ_{γ}, λ^{γ}, $V(h)$, and is of the order $\Lambda_N\left(\lambda^{\gamma}\right)$.

Note. A similar formula holds also for $\Lambda_N\left(\lambda^{\gamma}\right)$.

REFERENCES

1. Animalu A. Quantum Theory of Crystalline Solids, Moscow, Mir, 1981, 574 p.

2. Bethe G, Sommerfeld A. Internal Theory of Metals, Moscow, OGIZ, 1938, 316 p.

3. Bethe G, Sommerfeld A. Electron Theory of Metals, Moscow, Nauka, 1938, 532 p.

4. Berezin F., Shubin M. Schrodinger Equation, Moscow University Press, 1983, 388 p.

5. Blikmor J., "Solid state physics, Moscow, Metallurgiya, 1972, 451 p.

6. Veliev O.A. Asymptotic formulas for the eigenvalues of the multidimensional Schrödinger operator and periodic differential operators Preprint of Institute of Physics, Baku, № 157, 1985, 65 p.

7. Veliev O.A. Asymptotic formulas for the eigenvalues of the Schredinger operator and the Bethe-Sommerfeld conjecture, Functional analysis and applications, Vol.21, N.2, 1987, pp.1-5.

8. Veliev O.A. On the spectrum of multidimensional periodic operators, Theory of functions, functional analysis and applications, Issue 49, 1988, pp.17-34.

9. Vilenkin A.Y. Special functions and the theory of group representations, Moscow, Nauka, 1965, 252 p.

10. Gasymov M.G. Spectral analysis of a class of non-differential operators of second order, Functional analysis and applications, Vol. 34, N.2, 1980, pp.14-19.

11. Gelgand I.M. Eigenfunction expansion of the equation with periodic coefficients, Doklady Math., Vol. 73, N.6, 1950, pp.1117-1120.

12. Zayman J. The Principles of the Theory of Solids, Moscow, Mir, 1974, 472 p.

13. Zayman J. Electrons and Photons, "Moscow, Nauka, 1962, 512 p.

14. Kato T. Кото Т. Perturbation Theory of Linear Operators, Moscow, Mir, 1972, 740 p.

15. Kurant R. Partial Differential Equations, Moscow, Mir, 1964, 830 p.

16. Kittel Ch. Quantum Theory of Solids, Moscow, Nauka, 1967, 653 p.

17. Lazutkin V.F. Лазуткин В.Ф. Convex Billiards and Eigenfunctions of the Laplace Operator, LSU, 1981, 196 p.

18. Marchenko V.A. Марченко В.А. Sturm-Liouville Operators and Their Applications, Naukovo Dumka, Kiev, 1982, 231 p.

19. Maslov V.P. Asymptotic Methods and Perturbation Theory< Moscow, Nauka, 1988, 305 p.

20. Mehrabov V.A. On a spectrum of the periodic Pauli operator in the parallelogram (Bethe-Sommerfeld conjecture), News of Baku University, 2000, N.1, pp.110-124.

21. Mehrabov V.A. On the decomposition of the periodic Pauli operator on layers News of Baku University, 2000, №-4, pp.166-175.

22. Mehrabov V.A. Periodic Pauli operator. High order asymptotical formulas, Int. Conf. Modern problems of mathematics, mechanics and informatics, Tula, 2007, pp.61-63.

23. Mehraboov V.A. Asymptotic formulas for some series of non-resonance eigenvalues of three-dimensional periodic Pauli operator, News of Baku University, 2012, N.1, pp.38-48.

24. Naymark M.A. Linear Differential Equations, Moscow, Nauka, 1988, 305 p.

25. Popov V.N., Skriganov M.M. Note on the structure of the spectrum of the two-dimensional Schrödinger operator with a periodic potential, Notes of the seminar of LOMI, 1981, Vol.10, pp.131-133.

26. Rid M., Saymon B. Method of the Modern Mathematical Physics, Moscow. Mir, 1982, Vol.4, 428 p.

27. Riofe-Beketov F.S. Ф.С. On a spectrum of the nonselfadjoint differential operators with periodic coefficients, Doklady Math., Vol. 152, N.6, 1962, pp. 1312-1315.

28. Titchmarsh E.Ch. Decomposion over Eigenfunctions Releted to the Second Order Diffenential Equations, Moscow. IL, Vol.2, 1961, 556 p.

29. Harrison U. Theory of Solids, Moscow, Mir, 1972, 463 p.

30. Eastham M.S. The Spectral Theory of Periodic Differential Equations, Edinburgh, Scottish Academic Press, 1973, 130 p.

31. Rubinov J., Keller Y. Asymptotic solution of Dirac's equation, Phys. Rev., Vol.131, N. 6, 1963.

32. Thomas L.E. Time independent approach to scattering from impurities in a crystal, Comm. Math. Phys., Vol.33, 1973, p. 335-343.

Printed by Books on Demand GmbH, Norderstedt / Germany